A Geography of the European Union

A Regional and Economic Perspective

GARRETT **N**AGLE AND
KRIS **S**PENCER

OXFORD UNIVERSITY PRESS   1996

OXFORD
UNIVERSITY PRESS

Great Clarendon Street, Oxford OX2 6DP

Oxford University Press is a department of the University of Oxford.
It furthers the University's objective of excellence in research, scholarship,
and education by publishing worldwide in

Oxford New York

Athens Auckland Bangkok Bogotá Buenos Aires Calcutta
Cape Town Chennai Dar es Salaam Delhi Florence Hong Kong Istanbul
Karachi Kuala Lumpur Madrid Melbourne Mexico City Mumbai Nairobi
Paris São Paulo Singapore Taipei Tokyo Toronto Warsaw

with associated companies in Berlin Ibadan

Oxford is a registered trade mark of Oxford University Press
in the UK and in certain other countries

© Oxford University Press 1996

First published 1996
Reprinted 2000

ISBN 0 19 914655 1

All rights reserved. No part of this publication may be reproduced, stored in a retrieval system, or transmitted, in any form or by any means, without the prior permission in writting of Oxford University Press. Within the UK, exceptions are allowed in respect of any fair dealing for the purpose of research or private study, or criticism or review, as permitted under the Copyright, Designs, and Patents Act, 1988, or in the case of reprographic reproduction in accordance with the terms of licences issued by the Copyright Licensing Agency. Enquiries concerning reproduction outside those terms and in other countries sould be sent to the Rights Department, Oxford Unviersity Press, at the address above.

Typeset and designed by **Matthew Corbett Handy**

Artwork by **Hardlines, Charlbury, Oxford**

Printed in Hong Kong

The publisher and authors would like to thank the following for permission to reproduce photographs:

Aerocamera Rotterdam: pp. 99, 108; **Barnaby's Picture Library**: p. 110; **John Birdsall Photography**: pp. 104, 115; **Britstock/IFA**: p. 1; **Cardiff Bay Development Corporation**: p. 145; **Chorley & Handford Ltd**: pp. 50, 64; **Environmental Picture Library/Graham Burns**: p. 142; **Ford (UK) Ltd**: p 59; **Robert Harding Picture Library/Rolf Richardson**: p. 8; **Robert Harding Picture Library/Adam Woolfitt**: pp. 3 & 81; **Holt Studios International Ltd/Nigel Cattlin**: pp. 22 bottom left, 23; **IBM Scotland**: pp. 41 & 141; **Images Colour Library**: pp. 103, 137; **Jerrican/Berenquier**: p. 100; **Jaguar Cars Ltd**: p. 123; **Kommunalverband Ruhrgebiet**: pp. 63, 121; **Newcastle Central Library**: p. 70; **North News/T Ditchburn**: p. 71; **Northern Ireland Tourist Board**: p. 136; **OPW (Office of Public Works, Dublin)**: p. 138; **Oxford Scientific Films/GA Maclean**: p. 32; **Powerstock**: pp. 89 top, 90; **Rex Features Ltd**: p. 111 bottom; **Rex Features Ltd/Leon Schadeberg**: p. 93; **Rex Features Ltd/John Stephen**: p. 96; **Telegraph Colour Library**: p. 111 top; **Telegraph Colour Library/J Miles**: p. 89 bottom.

All other photographs are courtesy of **Garrett Nagle**.

Front cover photograph of Maastricht, Netherlands, by **Denis Waugh/Tony Stone Images**.

Every effort has been made to trace and contact copyright holders of material reproduced in this book. Any omissions will be rectified in subsequent printings if notice is given to the publisher.

The authors wish to express their deep gratitude to the many people, too numerous to mention, who helped at various stages along the production line. In particular, we wish to thank the Geography staff and students at St. Edward's, St. Paul's Girls and Westminster schools, for reading, using and commenting on early drafts of the book; Gill Yeomans; Misia Newsome; Claudia Harrison, Rachel and Kathy; Professor David Smith for the section on spatial margins; the anonymous referees for many useful comments; Matthew Corbett Handy for designing the layout and setting the text; Rosie and Patrick for sharing the midnight oil and Angela for support and encouragement.

For Agnes and Dan, Muriel and Ron.

Contents

Introduction IV

CHAPTER 1 CHANGE IN THE EUROPEAN UNION 1
    A Europe in the late twentieth century 2
    B Economic development and regional inequalities 6
    C Europe in a global context 12

CHAPTER 2 THE PRIMARY INDUSTRIES 19
    A Agriculture 20
    B The fishing industry 36

CHAPTER 3 FACTORS AFFECTING ECONOMIC AND REGIONAL DEVELOPMENT 41
    A Industrial change 42
    B Classical location theory 50
    C New location theory 52

CHAPTER 4 THE MATURE INDUSTRIES 59
    A Coal mining 60
    B Iron and steel 64
    C Textiles 68
    D Shipbuilding 70
    E Vehicles 74

CHAPTER 5 HIGH-TECHNOLOGY AND SERVICE INDUSTRIES 81
    A High-technology and knowledge-based industries 82
    B Service industries 88

CHAPTER 6 THE CORE 99
    A World cities 100
    B The Randstad 106

CHAPTER 7 REGIONS OF CHANGE 115
    A The UK's North-South divide 116
    B The West Midlands 120
    C Italy and regional development 122
    D The Ruhr 128

CHAPTER 8 THE PERIPHERY 133
    A Northern Ireland and the Republic of Ireland 134
    B Scotland and Wales 140
    C Greece, Portugal and Spain 146

Glossary 153
Index 154

'A Geography of the European Union' has been written with special reference to courses dealing with regional and economic geography. It contains a large number of new case studies and introduces new themes and concepts. It also reassesses some of the established thinking on geographic models and regional issues. Regions examined include the UK's West Midlands, Italy's Mezzogiorno, the Ruhr and the Randstad. The book also evaluates regional issues in less familiar areas such as Ireland, Northern Ireland, Spain and Portugal. Each chapter ends with a section of exercises and recommended reading. The exercises can be used to review the reader's understanding and are also structured to encourage a development of ideas and themes.

The book presents a detailed overview of the European Union which we think will prove of great value to a wide range of students.

Garrett Nagle and Kris Spencer

CHAPTER 1
CHANGE IN THE EUROPEAN UNION

A **EUROPE IN THE LATE TWENTIETH CENTURY**
  The Single Market 2
  Issues in the 'New Europe' 4

B **ECONOMIC DEVELOPMENT AND REGIONAL INEQUALITIES**
  Measuring regional inequalities 6
  Problem regions 8

C **EUROPE IN A GLOBAL CONTEXT**
  Trade relations 12
  Global shift 14
  Competitiveness 15

D EXERCISES AND RECOMMENDED READING 18

In **Section A** the most important trends in Europe's development are examined. These include demographic, economic and political forces which have evolved and changed over the last two centuries. Although the fortunes of some areas and sectors have fluctuated considerably, there is now an emerging core, uniquely positioned and eager to exploit the 'new' Europe's role in the modern world system.

Section B analyses regional inequalities in Europe and investigates the way in which countries and regions develop. This starts with a simple model illustrating the transition from a primitive agricultural society to a complex urban-industrial one (Clark's sector theory) and then examines how this change may be brought about (Rostow's model of economic development). To understand why certain countries, and indeed certain regions, prosper while others falter, Myrdal's model of cumulative causation is assessed and considered in relation to the 'new' Europe. The 'best location' in Europe is illustrated by the rapidly developing 'hot banana', ranging from northern Italy to the South East region of the United Kingdom.

Section C looks in detail at the changing status of Europe in the new global system. This section starts with a discussion of the EU's trading relations with the USA, Japan and the developing world. It then considers the global shift of manufacturing towards the South East Asian economies. This leads on to a discussion of competitiveness and productivity. The UK is given as a case study of the complexities of foreign direct investment. These raise a number of issues—social, economic, political and environmental—which are developed in detail throughout subsequent chapters of the book.

Section A Europe in the late twentieth century

The European Union (EU) of 15 European states has developed from its beginnings with the Treaty of Paris (1952) to become the most powerful trading bloc in the world. It has a combined population of 370 million, producing goods and services valued at over ECU 3,000 billion (ECU 1 = £0.75). Its evolution has been unsteady and occasionally acrimonious. With the accession of each additional country a new set of economic and political problems has been added: from the insularity of Britain to the peripherality of Portugal and Greece. Between 1952 and 1973 there were only six member states—Belgium, France, Germany, Italy, Luxembourg and the Netherlands. In 1973, Denmark, Ireland and the UK joined. They were followed by Greece in 1981 and Spain and Portugal in 1986. The three latest countries to become members were Sweden, Finland and Austria who joined on 1 January 1995 (Figure 1.1). Entry to the EU is open to all democratic, market-oriented European states which means that Turks, Cypriots, Albanians and Russians are potential members if they adopt suitable economic, political, judicial and constitutional systems (Inset 1.1).

THE SINGLE MARKET

The ratification of the **Single European Act** in 1986 led to the disappearance of most frontier controls by the beginning of 1993. The legislation gives the EU a marketplace which is protected by its unity against attacks from Japan, the USA and increasingly the newly-industrialising countries of the Pacific Rim. This idea of 'Fortress Europe' will be discussed in **Section C** of this chapter.

Within the EU the Single Market has increased competition between the member states by effectively removing national boundaries. In essence, each member state is free to trade within the EU without fear of trade restrictions. Regionally, it is hoped the Act will lead to convergence: that is, capital and labour disparities in incomes and profitability will be narrowed between the richer core and poorer periphery (**Figure 1.2**).

However, the single market has had some harsh effects. Plants have closed and companies have laid off workers as industry has attempted to rationalise operations in the most productive or lowest cost centres. Another problem is that 1993 marked the 'coming of age' of many of the EU's peripheral members such as Spain, Portugal and Greece. Until this point they had benefited from huge influxes of Structural Fund money while being largely protected from other member states. The large frontier-free market created by the Single European Act could have disastrous consequences for the economies of states and regions with problems of inefficient agriculture, out-dated industries or poorly developed infrastructure.

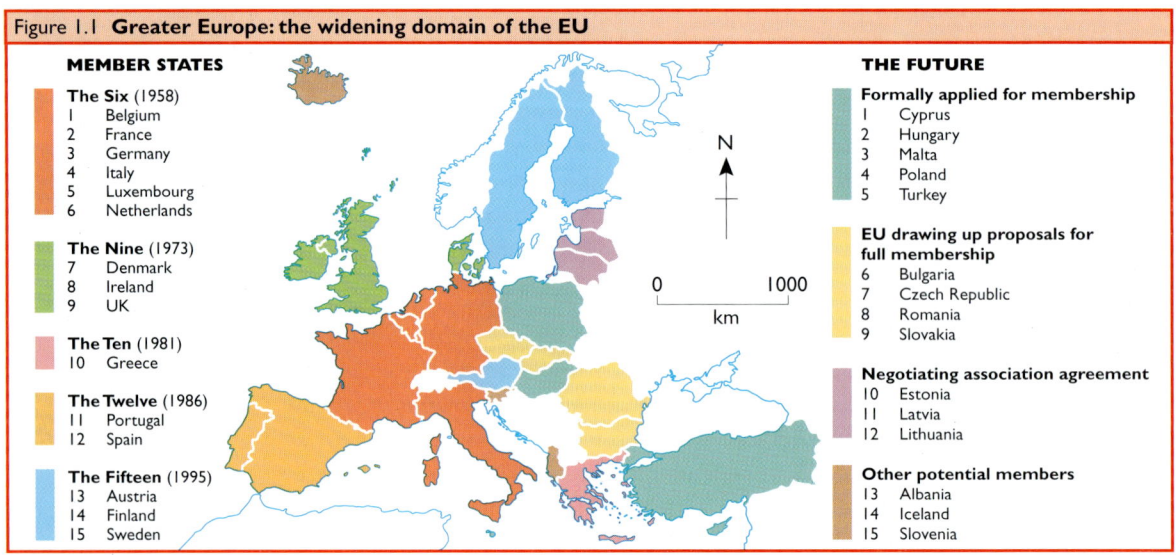

Figure 1.1 **Greater Europe: the widening domain of the EU**

Figure 1.2 Europe's core and periphery

INSET 1.1 **ENLARGING THE UNION**

The EU faces a number of regional forces from within Europe which it cannot ignore:

1. *Reunification of Germany has shifted the regional balance towards the east and may foreshadow an expansion of the EU to include the Czech and Slovak states, Poland and Hungary.*

2. *There is also pressure from Western Europe. In 1991 EFTA — the European Free Trade Agreement between Austria, Finland, Iceland, Liechtenstein, Norway, Sweden and Switzerland — formed the European Economic Area (EEA) with the EU uniting a population of 370 million people.*

3. *Cyprus, Malta and Turkey have been petitioning for inclusion for many years but if they joined structural funding would be severely overstretched.*

The new members from EFTA will shift the Union's geographic centre north and east, towards Germany, so further peripheralising countries like Spain and Portugal.

In addition to the economic issues of further enlarging the Union to 16 or 20 member states there are political ramifications. Enlargement of the EU to the east is unlikely in the short-term but the central and eastern European countries are pushing for improved access to the market of the EU. The goal is clearly political stability in the east but even the most developed regions — Poland, Hungary, the Czech Republic and Slovakia: the 'Visegrad countries' — have only one-third per capita income of the fifteen and are twice as dependent on agriculture. It is unlikely that the EU could fund their convergence. Moreover, there is little agreement between EU countries as regards the benefits of enlargement; peripheral countries in particular fear they may lose regional assistance.

ISSUES IN THE 'NEW EUROPE'

Europe's task as it enters the next century is to show it can face up to a number of problems and opportunities. There is little disagreement on the diagnosis of Europe's problems, the task is finding solutions which are complementary. The EU must address the following weaknesses:

1. Growth of **public spending** may well outstrip the EU's resources as its population ages, leading to a higher dependency ratio.

2. Long-term **unemployment** means that more flexible markets are needed which could include an increase in part-time work, adults with more than one job or an integrated tax and welfare system.

3. The EU needs to improve its **technological base** and the skills and flexibility of its **workforce**.

4. A balance must be found between the costs and benefits of **environmental regulation**.

In effect, Europe has a number of areas of concern which it shares with other developed countries: unemployment, sectoral change, environmental issues and its ageing population.

Employment

According to the 1995 *Regional Trends*, in December 1992, 9% of EU citizens were without a job. The rate of increase in unemployment was highest in Spain (17.8%), Ireland (15.3%), France (10.0%) and the UK (9.8%).

The decline of labour-intensive industries suggests that even renewed economic growth cannot generate the kind of job creation needed. Almost half of all those unemployed in Europe in 1990 had been out of work for a year or longer and one-third for at least two years. This is especially problematic in peripheral regions. In Italy, for example, 78% of the long term unemployed have never worked at all. The phenomenon of youth unemployment is particularly marked in Spain and Italy where the level is close to 30% (**Figure 1.3**).

There are two main theories put forward to explain Europe's present joblessness:

- The EU's unemployment is part of an **economic cycle**.
- The EU's unemployment is mainly **structural** and requires improved competitiveness.

Others agree that unemployment is structural but would raise **import barriers** against suppliers in Eastern Europe and the NICs. It is clear that long-term unemployment will remain in the EU and whatever the cause, Europe may have to change the way it views 'employment' and education. It has been suggested that the labour market should have the flexibility to move workers displaced from contracting industries into new jobs in expanding ones. The European Single Market should encourage the creation of new jobs but the geographic distribution of these jobs is unlikely to be evenly spread. Indeed, the relative immobility of the EU's workforce remains on important barrier to economic integration.

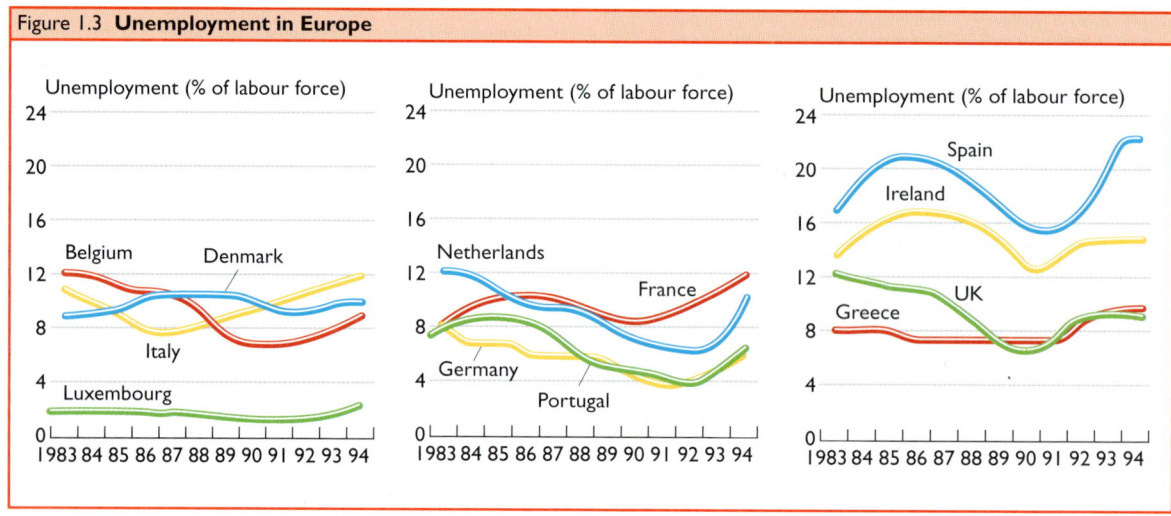

Figure 1.3 **Unemployment in Europe**

Sectoral change

The rise of **long term unemployment** in Europe is linked to a decline in the importance of manufacturing as an employer and as a contributor to GDP. The service sector is Europe's largest source of output and employment—and its importance is increasing. In the former West Germany, Europe's leading manufacturing economy, the number of jobs in services rose by 28% between 1990 and 1992, while totals in other sectors stagnated or declined.

The outlook is for a continuing fall in factory jobs and a squeeze in public sector payrolls throughout Europe, leaving services as one of the few areas of job creation. However, the search for new jobs coincides awkwardly with structural upheavals in services businesses such as telecommunications, the media, financial services and retailing. **Rationalisation** and the drive for efficiency means that Europe cannot even depend on these businesses for job creation. The Single Market looks set to increase cross-border competition. Desk jobs look less secure in the light of labour-saving technologies and privatisation which are challenging bureaucracies.

Ageing Europe

The labour market may well tighten as the number of people of working age shrinks due to Europe's ageing population. In the 18 Western European member states of the OECD the number of people aged 65 and over will rise from 50 million to more than 70 million between 1990 and 2030. During the same period, the number of people of working age will fall. The result is that by 2030 there will be fewer than three people of working age in these countries for each person over 65, compared with five now. If this **demographic restructuring** eases some employment problems it is likely to have more serious effects on patterns of consumption, production, savings, investment and politics. The greatest impact will be felt in Germany, where the population will decline by about 15 million between now and 2030. Today, a fifth of the population is 20 or younger and another fifth 60 or over. By 2030 only 16% will be under 20, 46% will be between 20 and 60, and 30% over 60.

Environment

In the past two decades European countries have created some 200 environmental rules. Many European countries are concerned that these could cause some manufacturers to leave the EU with resultant job losses. According to the OECD, total environmental spending by public and private sectors in northern European countries appear to outstrip that of Canada and Japan, and to be roughly equivalent to that in the USA—around 1.4% to 1.6% of GDP (**Figure 1.4**). The issues are as much global as European:

1. Northern European companies claim that they face tougher rules than companies in southern Europe. German spending, the highest in Europe, is roughly double that of Italy, Portugal and Spain.

2. Many feel the strict laws in northern Europe leads to 'environmental dumping' of dirty industries into the peripheral south or even outside the EU.

3. Germany's inheritance of the environmental problems of the old East Germany further complicate the problem.

The end result may be that Europe loses companies and so jobs. For ICI, the international chemical group, the need to comply with mercury emission standards led to the closing of the chlorine plant at Hillhouse in the UK.

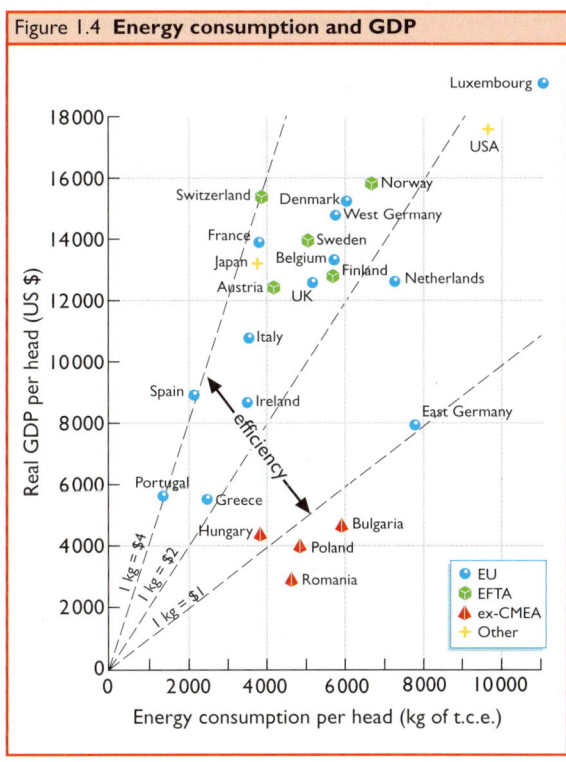

Figure 1.4 **Energy consumption and GDP**

Section B Economic development and regional inequalities

MEASURING REGIONAL INEQUALITIES

A variety of economic, social, demographic and environmental indices can be used to measure regional variations in development and standard of living: GDP, unemployment rates, labour costs, income per head, proportion of GDP from agriculture, proportion of population employed in agriculture and declining industries, type of industry in a region, receipt of regional and structural aid, investment into the area, age-structure of the population, rates of out-migration, number of people in further education, housing quality, social service provision and environmental dereliction. However, these are very difficult to quantify, and at best offer only a partial description. The use of any single measure has many difficulties and a combination of indices is therefore necessary for a more realistic picture.

Gross domestic product (GDP)

Figure 1.5 shows variations in the level of GDP throughout the regions of the EU. The highest levels, above 150% of the EU average, are found in the small city-regions, such as Brussels, Bremen, Hamburg and the Ille de France. These represent Europe's inner economic **core**. An arc of regions from the South East of the UK, through north and west Netherlands, northern Belgium, much of the former West Germany and into northern Italy forms a secondary or outer core, with GDP levels of between 110 and 149% of the average. The **inner periphery** is characterised by low levels of GDP: Northern Ireland, Ireland, Spain, Portugal, southern Italy and the Attiki region of Greece receive between 80% and 50% of the EU average. The lowest GDP, less than 50% of the EU average, is

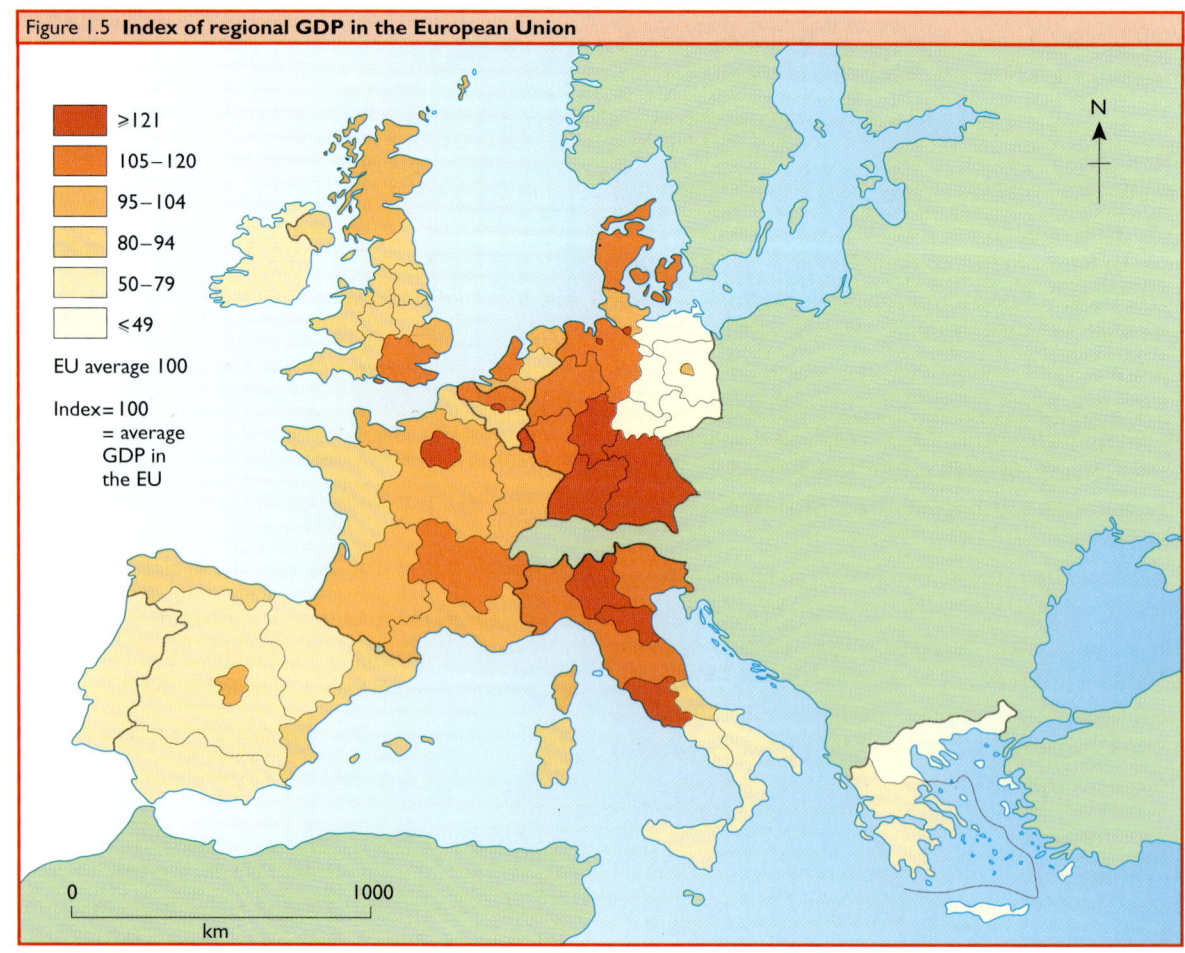

Figure 1.5 **Index of regional GDP in the European Union**

found in the rest of Greece and the former East Germany excepting Berlin. These signify the **outer periphery**. Thus a clear core-periphery can be identified on the basis of GDP.

Unemployment

The pattern for unemployment is almost the reverse of that for GDP and shows clearly a core of low unemployment and a periphery of high unemployment (**Figure 1.3, p.4**). Highest rates (>15%) are experienced in Spain, Ireland and southern Italy. These areas are relatively **rural** and have lacked industrialisation. By contrast, moderately high rates of between 13% and 15% are found in central Italy, Northern Ireland, the Highlands of Scotland, Greater London, Birmingham, northern France and Belgium, and are associated with a mixture of **counter-urbanisation**, **deindustrialisation** and a **lack of industrialisation**. Low rates, less than 7.8%, characterise Germany, northern Italy, and Greece, while the lowest rates of unemployment, less than 5.2%, are seen in southern Germany, Luxembourg and, surprisingly, Portugal.

Opportunity cost

All locations offer opportunities, such as access to labour, raw materials and transport infrastructure, which directly affect the cost of running a business or organisation. However, by locating in one area, the opportunities or advantages of other areas are lost—this is known as **opportunity cost**. It would be wrong to suggest that to be peripheral within the EU is wholly negative. Peripherality means that a country or region is eligible for structural funds to promote economic convergence, and there is also the obvious advantage of lower operating costs which are usually to be found in the peripheral regions.

Three areas emerge from a consideration of operating costs:

1 **High cost countries**: (West) Germany, parts of France and the Benelux countries.

2 **Intermediate**: the UK, much of France and Italy, with lower costs than the core but the advantage over the periphery of a better developed infrastructure, a skilled workforce and technological sophistication.

3 **Low-cost non-central**: Ireland, the Iberian Peninsula and Greece.

The best location in Europe

Any discussion of the locational advantages of regions within Europe obviously begs the question 'Where is the best location in Europe?' An **assembly plant** may locate in an area of cheap labour or where financial assistance is available: this would favour a peripheral area. Another company, developing an **R&D centre**, would look for skilled links to a university or related research companies. Furthermore, a company from outside Europe may be welcomed by one country but shunned by another. The UK's open door policy, coupled with a relatively inexpensive but skilled labour force, has enabled it to attract 40% of all US and Japanese inward investment in the EU.

What is emerging from this discussion is a number of ways of thinking about success and attractiveness in the regions of Europe. The 'hot banana' is a model based largely on accessibility and centrality. Other key factors in the decision-making process include good transport links, modern telecommunications, cost differentials and a supportive business climate.

1 **Transport**: effective, efficient transport links are universally regarded as essential and while recent huge investment projects are being completed around the EU—the Channel Tunnel, the Rhine-Danube canal and a community-wide network of high-speed trains—many regions are still poorly linked to the heart of Europe.

2 **Labour**: for companies with labour-intensive activities the cost and skill of labour is often the most important locational factor, hence the regions of southern Europe and Ireland offer the most attractive labour force, although local skill levels, labour laws and industrial relations may be less attractive.

3 **Taxation and financial assistance**: national systems of taxation differ markedly, with Germany imposing the lowest tax burden on corporations and France, Portugal and Spain the highest. On the other hand, all countries offer a variety of packages to attract investment and these are in addition to EU measures.

PROBLEM REGIONS

From the examples above it is possible to identify a number of types of problem region. These include:

1. **Depressed rural areas** are characterised by below average incomes, limited industrialisation, agriculture as a declining source of livelihood and serious out-migration. For example, the Mezzogiorno in southern Italy remains geographically peripheral and economically disadvantaged. Progress has occurred but there is still serious out-migration, low rates of income and relatively little investment. The EU defines problem regions in terms of eligibility for financial incentives. Those areas with a GDP of less than 25% of the EU average are classified as **Objective 1**: this includes Ireland, Greece, Portugal, over half of Spain and over one-third of Italy. These are generally the depressed rural areas.

2. **Declining industrial areas** include the Thames Gateway, the Nord-Pas-de-Calais and the North East of England. Associated with with early industrialisation, these areas now suffer from deindustrialisation: the decline of traditional industries such as textiles, shipbuilding and iron and steel manufacturing. Often there is a low standard of living, poor environmental quality and a lack of economic growth. Industrial readjustment is needed but is slow and difficult. According to the EU these are **Objective 2** areas. Most of the areas covered are in the traditional industrial areas: more than one-third of the population in the UK, Belgium and Luxembourg is covered. Smaller areas include Nord-Pas-de-Calais and the Basque region of Spain. Although the Ruhr is classified as an Objective 2 area, it is an excellent example of a region which is beginning to shake off its declining status, largely due to its ambitious economic and environmental readjustment programme.

3. **Primate regions** are associated with an over-concentration of wealth, opportunity, industry, R&D, population and political power in urban and industrial areas. Examples include London, Paris and the Randstad of the Netherlands. This gross inequality in opportunity was most evident in France and was labelled 'Paris and the French desert' by Gravier in 1957.

It is equally possible to delimit a number of growth regions. These can be categorised as:

1. **The Golden Triangle** bounded by Paris (Ille de France), London (the Home Counties) and Amsterdam (Noord-Holland) and including the Ruhr Basin, which is surrounded by other regions which share good accessibility, skilled labour and technological sophistication.

2. **Islands of innovation**, additional regions within core countries such as Berlin, Toulouse and Bordeaux specialising in techno-industrial fields.

3. **Activity centres** within peripheral areas such as parts of Ireland and Scotland, the Lisbon-Oporto axis in Portugal, the Spanish Mediterranean coast (Barcelona, Valencia and Malaga) and the Mezzogiorno (Bari) in Italy which have attracted inward investment through low labour costs and financial incentives.

Figure 1.6 **Canary Wharf: part of the EU's core**

ECONOMIC DEVELOPMENT AND REGIONAL INEQUALITIES

INSET 1.2 THE 'HOT BANANA'

EU policy makers have drawn up a map of Europe's core areas based largely on market accessibility. The result is a banana-shaped zone which stretches from the South East of the UK, through the Benelux countries, northern France, the Rhine and Ruhr, to Milan and northern Italy. This zone, the so-called 'hot banana', is likely to attract head-offices and research facilities as a result of superior transport and telecommunications. Indeed, the zone accounts for only 10% of the surface area of the European Economic Area (EU and EFTA nations combined) but for more than 40% of its total output. The peripheral areas include the Scandinavian countries, Greece, southern Italy, Ireland and Scotland. With 60% of the EEA land area they account for only 20% of total output.

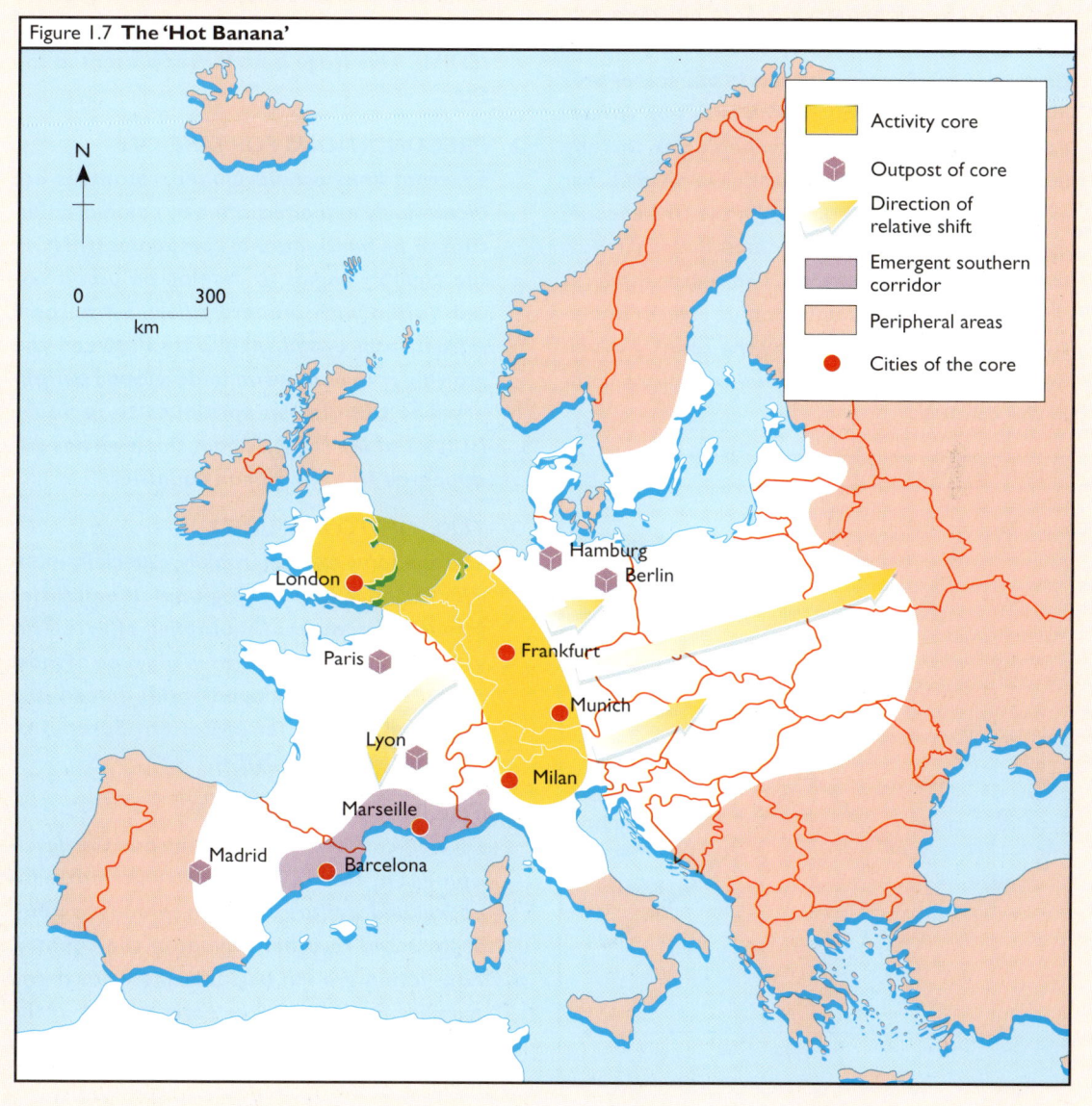

Figure 1.7 The 'Hot Banana'

INSET 1.3 MODELS OF DEVELOPMENT

CLARK'S SECTOR MODEL

All countries have progressed from agricultural societies to industrial and service economies. For some, such as the UK and Germany, the transition was mostly in the nineteenth century whereas for others, such as Ireland and Portugal, it occurred during the twentieth century. The model clearly shows the transition from an economy dominated by the primary sector to one dominated in turn by the secondary and tertiary sector. Change occurs because success in one sector produces a surplus revenue which is in turn invested into new industries and technologies, thereby increasing the range of industries in an area. For example, the cotton industry (in the UK) encouraged textile machinery, other metallurgical industries and service industries. The sector model is descriptive and offers only a crude level of analysis. It does not say how or why the country developed, nor does it show internal variations within the country.

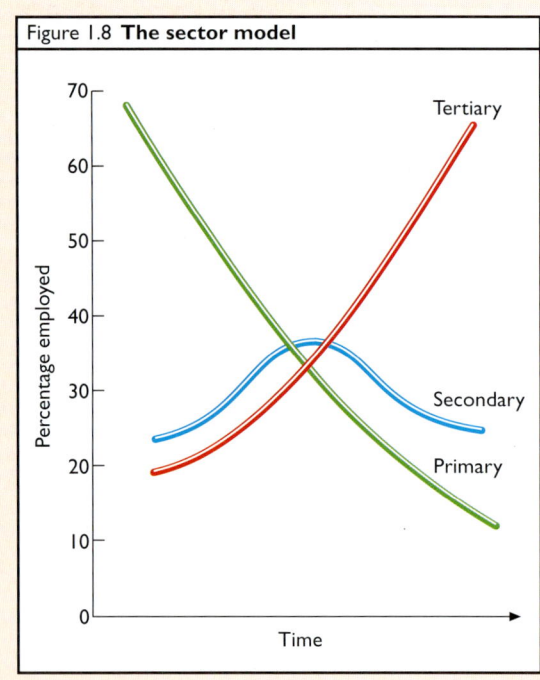

Figure 1.8 The sector model

ROSTOW'S MODEL OF DEVELOPMENT

W. W. Rostow, a US economist, envisaged five stages in the development of an economy. His model is a useful starting point in describing and understanding levels of development.

1. **TRADITIONAL SUBSISTENCE ECONOMY**
   Agricultural basis, little industry, few external links and low levels of population growth. Stage 1 of the demographic transition model (DTM). This stage is no longer present in the EU.

2. **PRECONDITIONS FOR TAKE-OFF**
   External links are developed; resources are increasingly exploited, often by colonial countries or by multinational companies (MNCs); the country begins to develop an urban system (often with primate cities), a transport infrastructure and inequality between the growing core and the underdeveloped periphery. The population continues to increase (stage 2 of the DTM). Again, this level has disappeared from European countries.

3. **TAKE-OFF TO MATURITY**
   The economy expands rapidly, especially manufacturing exports. Regional inequalities intensify because of multiplier effects. This growth can be 'natural' (as in the case of most countries of the developed world), 'forced' (the ex-socialist countries of Eastern Europe) or 'planned' (as in the NICs).

4. **THE DRIVE TO MATURITY**
   Diversification of the economy and the development of the service industry (health, education and welfare). Growth spreads to other sectors and to other regions in the country. Population growth begins to slow down and stabilise (late stage 3 or early stage 4 of the DTM). Ireland, Greece, Spain and Portugal are at this level.

5 THE AGE OF HIGH MASS CONSUMPTION
Advanced urban-industrial systems, with high production and consumption of consumer goods, such as televisions, compact disc players and dishwashers. Population growth slows considerably (stage 4 of the DTM). The UK and Germany characterise this level.

The main weaknesses of Rostow's model are:
1 It is based on the experience of countries in Western Europe and North America.
2 It is aspatial and does not look at variations within a country. For example, within the UK or Italy there are great disparities in the levels of development between the north and south of the countries — Rostow's model fails to pick this out.

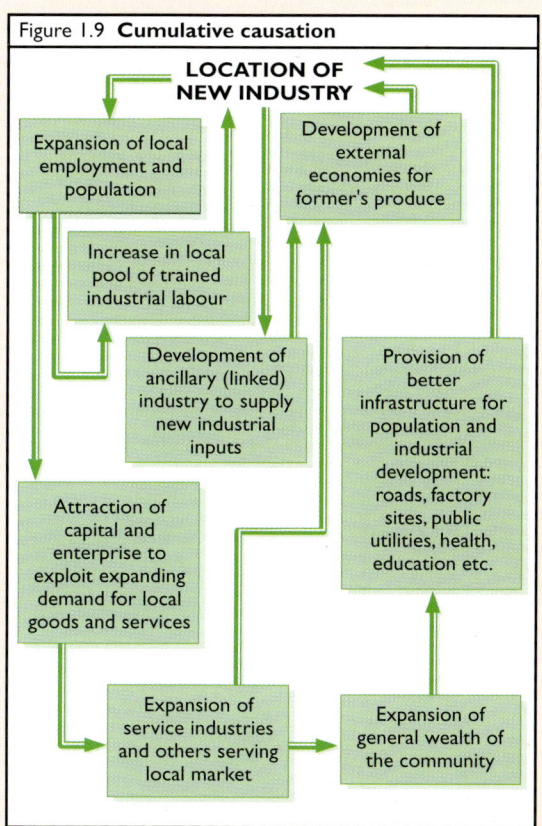

Figure 1.9 **Cumulative causation**

MYRDAL'S MODEL OF DEVELOPMENT

In 'Rich Lands and Poor Lands' (1957) Gunnar Myrdal argued that, over time, economic forces increase regional inequalities rather than reduce them. Development had two causes:

1 CUMULATIVE CAUSATION
Initial comparative advantages, e.g. natural resources or labour, create the initial stimulus for an industry to develop. In turn, cumulative causation ('multiplier effect') occurs as acquired advantages, such as infrastructure improvements, skilled workforce and increased tax revenues, are developed and reinforce the area's reputation, attracting more investment, ensuring growth and regional dominance.

2 SPATIAL INTERACTION
This increases growth in the core, while the peripheral areas are inundated by manufactured goods from the core (the 'backwash effect'), thus preventing the development of a local manufacturing base. As the core expands it may stimulate surrounding areas because of increased consumer demand: the 'spread effect'.

Three stages can be identified in Myrdal's model:
1 Traditional, pre-industrial stage with few regional disparities (Rostow's stage 1).
2 Increased disparities caused by multiplier and backwash effects as the country industrialises (Rostow's stages 2 and 3).
3 A reduction in regional inequalities as spread effects occur (Rostow's stages 4 and 5).

Myrdal's ideas have been used extensively in regional planning. For example in 'growth pole' policies, places or districts are developed by planners to form natural growth poles and these expand faster than other districts. Generally they are urban-industrial complexes which are economically more attractive due to natural resources, labour or accessibility, e.g. Dunkirk and Taranto.

Section C Europe in a global context

Sections A and B discussed issues within the European Union. However, the EU also faces challenges from the rest of the world. It must adapt to cope with the collapse of the Soviet Union, the admission of new members and eventual expansion to digest the new democracies of central Europe. At the same time, it must compete and bargain with the Japanese and the newly resurgent USA. It must also face the growing economic strength of the newly-industrialising countries (NICs). At this stage it may be useful to identify the tensions and opportunities which exist between the EU and the rest of the world.

TRADE RELATIONS

The rapid growth in world trade since the 1950s has seen a globalisation of formerly separate economic and industrial systems. **Globalisation** is the international expansion and integration of key corporate functions such as research and development (R&D), production and marketing as well as the growth in international collaboration and networking with other firms. It also involves the growing unification and integration of industrial activities across countries (**Figure 1.10**) and the rise of trading blocs (**Inset 1.4**).

The result has been a mixture of advantages and disadvantages for the EU and its trading partners.

The United States of America and the EU

US–EU trade tensions are dominated by frictions over agriculture. The USA sees the subsidised European produce as competing unfairly both within the USA and abroad. Americans view the enlargement of an 'inward-looking' EU with suspicion. They believe that it is too early to judge whether the Single Market is a 'market opening' helping the EU develop into a non-discriminatory trading bloc or if it is the genesis of a protectionist **'Fortress Europe'**. However, US transplant firms have a strong presence in Europe, especially with respect to electronics and automobiles. The USA, which makes around 40% of all its European investments in the UK, could be looking elsewhere for new investment. The North American Free Trade Agreement (NAFTA) between the USA, Canada and Mexico means that US MNCs are being encouraged to locate closer to home.

Japan and the EU

Japan–EU trade tensions are dominated by disagreements over Japanese car exports to the EU. Japan has accepted temporary restrictions until 1999. The French and Italians are still aggrieved. The Japanese producers have **transplant factories** in the UK capable of producing over 1.2 million cars by 1999. These cars will be classed as 'European', due to local content sourcing, and so will not be subject to import restrictions. The fears that these are little more than 'screwdriver' assembly plants seem largely unfounded. The Japanese are drawn to producing a 'European' car, researched and managed from Europe. In return the plants offer jobs, regional regeneration and transfer of technology to the host country.

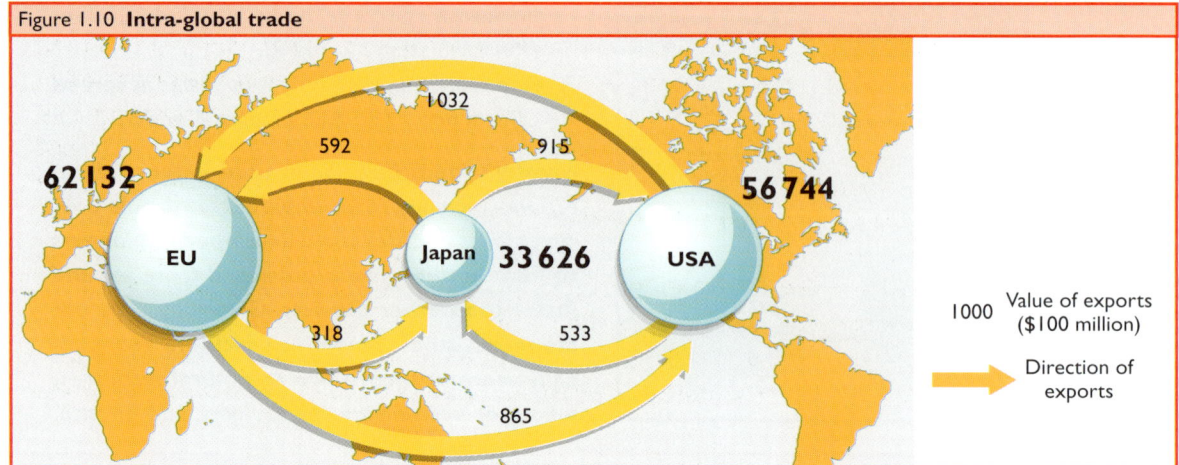

Figure 1.10 **Intra-global trade**

The developing world and the EU

Much of the increased growth generated by the Single Market will result from trade diversion, especially in manufacturing products through the elimination of internal customs and the standardisation of products. Imports from the rest of the world may fall by up to 10%. However, a recent Overseas Development Institute (ODI) survey put export gains for developing countries at about $10 billion, equivalent to 0.25% of the developing world's GNP. However, because about 80% of the gains will come from increased demand for fuel the gains to most countries will be small.

Sub-Saharan African and Latin American countries would have some benefit from liberalisation of primary production, while the Asian NICs are likely to gain most of the benefits of trade creation in manufactured products. Some developing countries will benefit at the expense of others. For example, the substitution of EU quotas for national quotas should be beneficial to efficient Latin American banana production.

INSET 1.4 TRADING BLOCS

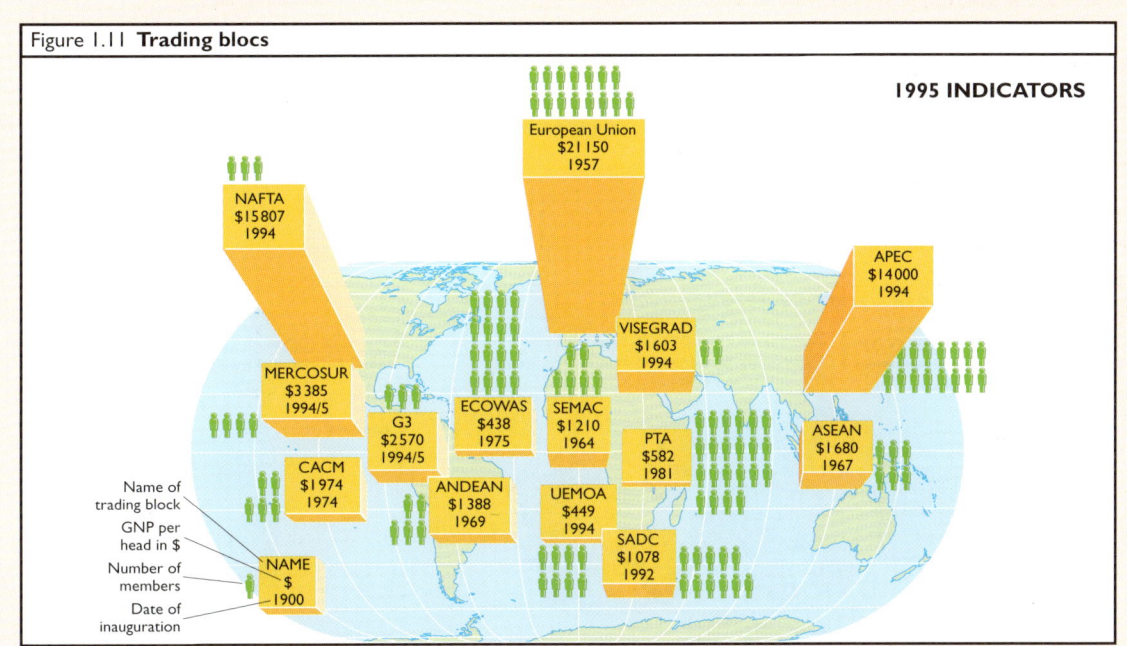

Figure 1.11 **Trading blocs**

There are two major types of player when it comes to world trade: the multinational corporation (MNC) and the regional trading bloc. In many ways both undermine the power of the single nation state. MNCs move freely between nations trading goods and information. Trading blocs are regional groups like the EU which defend their own industries and markets and sometimes impose protectionist measures on non-members. Between the two is the World Trade Organisation (WTO), which replaced GATT in 1995, and aims to encourage trade worldwide.

There is great variety amongst trading blocs. The EU is a unique example of 'deep' integration and has certainly encouraged free trade within the Union. However, some geographers fear that trade within regional blocs is at the expense of trade with the rest of the world. At the heart of this debate is the question of whether the EU is an organisation to promote trade and competitiveness or to protect its industries from the threat of imports and MNCs generated by overseas economies, above all, in east Asia.

GLOBAL SHIFT

One of the major changes the EU faces is a shift of economic power from the West to Far East. The four 'little dragons' of Asia—Singapore, South Korea, Taiwan and Hong Kong—have recorded GDP growth rates of between 3% and 5% over the 1990s. More importantly, as India and China come out of self-imposed exile, the world supply of semi-skilled labour has doubled and wage rates are falling. Both countries look set to emerge as major economic powers. China has had a real economic growth of 9% over the 1990s with an explosion of private enterprise. The liberalisation of India's economy has led to a growth rate of 5.4% a year in the period 1981–91, with considerable growth in high-technology R&D. The EU is not equipped for this competition:

- **Standards of education** in the NICs are equal to if not higher than those in the EU: 70% of South Koreans go on to university.
- **Productivity** is higher in the East.
- **Labour costs** are lower.
- **Social and economic expectations** are far lower.

Figure 1.12 shows the potential effects of this competition over the next 25 years. There will be a number of economic and social consequences of this new and increasing competition:

1. **Unemployment**: long-term unemployment will be the norm for a significant proportion of the adult population which could lead to social unrest, especially in inner city areas.

2. **Investment**: Europe must provide money for education in order to promote a flexible and multi-skilled workforce open to new technology.

3. **New opportunities**: half of the world's population do not have access to electricity and there are opportunities for European industry to build furnaces, insure them and run them; e.g. North West Water, in the UK, manages Malaysian Waste Management.

The clear message is that the countries and industries of the EU must become more competitive.

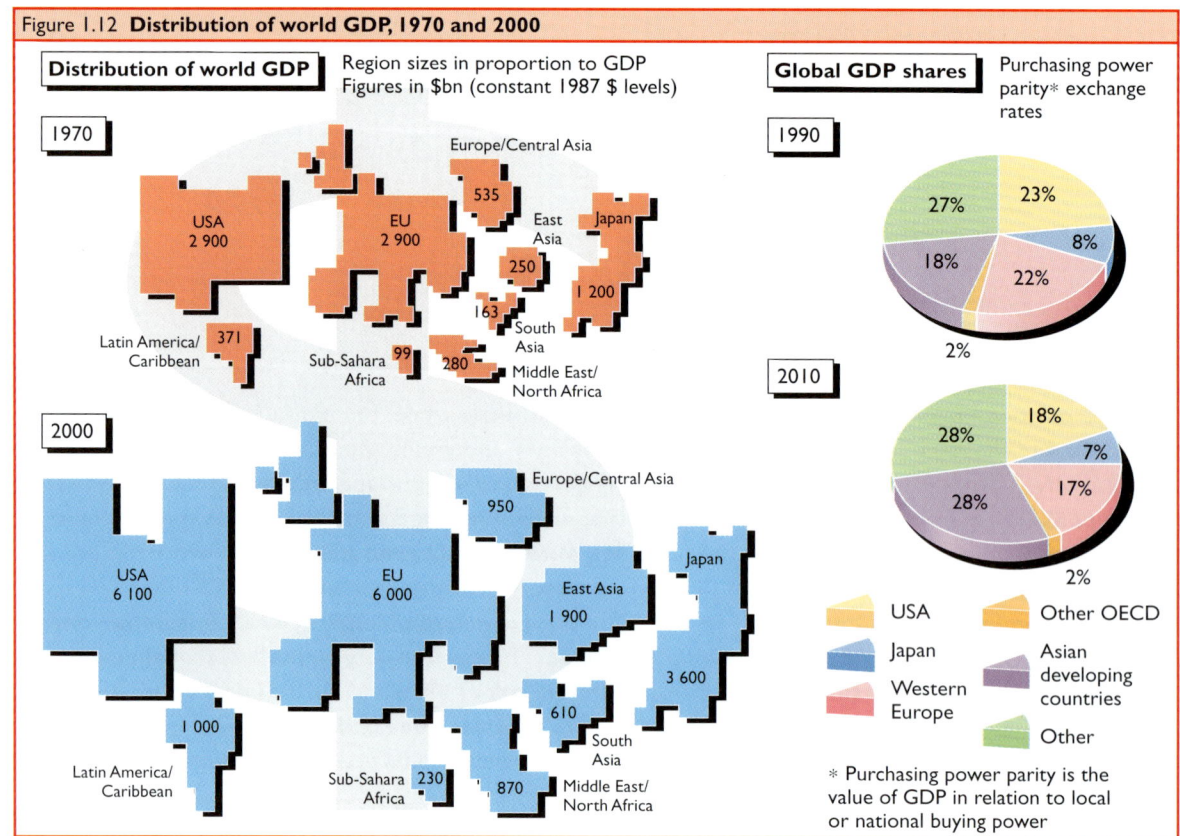

Figure 1.12 Distribution of world GDP, 1970 and 2000

COMPETITIVENESS

The concept of **competitiveness** is very much at the heart of this book. The comparison of regional and industrial performance at a variety of scales is the root of economic geography. The discussion of competitive advantage gives a clear indication of the present and future potential of region, country, trading bloc, small firm or multinational corporation.

Competitiveness has been defined by factors like growth of GDP, labour costs, trade balances and exchange rates. However, ability to compete also includes a number of more qualitative factors such as:

◆ Motivation, education and values.
◆ Labour skills and entrepreneurial flair.
◆ Flexibility of workforce and production.
◆ Interaction of politics, business and industrial organisations.

One of the most common measures of competitiveness is **productivity** which is measured as value added per manufacturing worker. Productivity can be measured at a variety of scales between regions and industries. **Figure 1.13** shows that Japan has the best ranking: each Japanese manufacturing worker produces over $14,000 more than a US worker and $21,000 more than the average EU worker.

Figure 1.14 shows a more general study of competitiveness based on the management practices and manufacturing performance of 71 automotive component plants in number of countries. The survey derived a number of interesting findings:

1. From a regional perspective the order of performance was Japan, followed by the USA and then Europe.
2. There were a number of 'world class' plants which simultaneously achieved high productivity and high quality.
3. In terms of quality (defects) Japan had a 4:1 advantage over France and Germany and an overwhelming advantage over the UK and Italy (at least 8:1).

Most industrial and regional policies aim to increase competitiveness in some way. At the European scale the result is a complex web of policies which aim to improve the competitiveness of peripheral regions within the EU and also promote its industries at a global scale. As we will see, this drive for European competitiveness is not without problems; what may help one industry or region often acts against another.

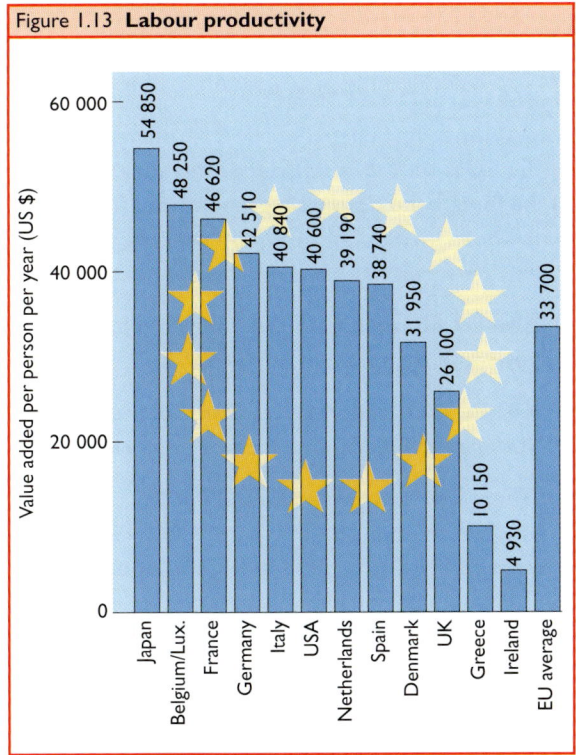

Figure 1.13 Labour productivity

Figure 1.14 Global manufacturing competitiveness

Countries	Japan	Spain	France	USA	UK	Italy	Germany
No. of world-class plants	5	2	3	3	-	-	-
Productivity	●	●	◐	○	◐	◐	○
Labour costs per unit	◐	●	●	◐	○	○	●
Incoming defects	●	●	◐	○	○	○	◐
Internal defects	●	●	◐	◐	◐	○	○
Customer complaints (ppm)	●	◐	◐	●	○	○	◐
Stock turns	●	◐	○	●	◐	◐	○

Key
● ranked 1 or 2   ◐ ranked 3, 4 or 5   ○ ranked 6 or 7

INSET 1.5 INVESTMENT, THE UK AND THE EU

The 1990s have seen a new word enter into the British vocabulary: 'Euro-scepticism'. There is an unease about the amount which the UK contributes to the budget of the EU. At the same time, issues such as a single European currency and the 'federalisation' of Europe are seen to call Britain's sovereignty and self-determination into question. In addition to these political issues are a set of economic questions about the UK's membership of the EU. Figure 1.18 shows that the economic argument is finely balanced. One important issue is the effect that membership has on the flow of capital investment both into and out from the UK.

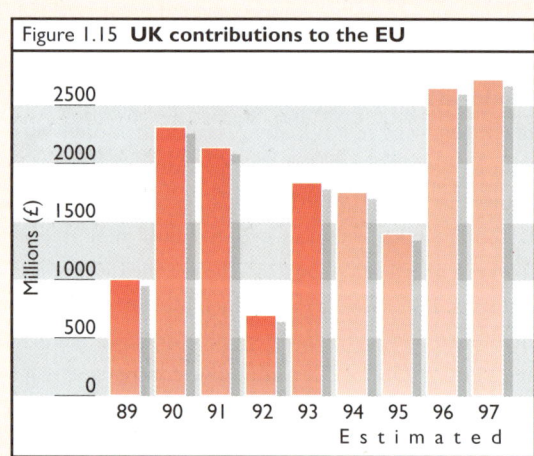

Figure 1.15 UK contributions to the EU

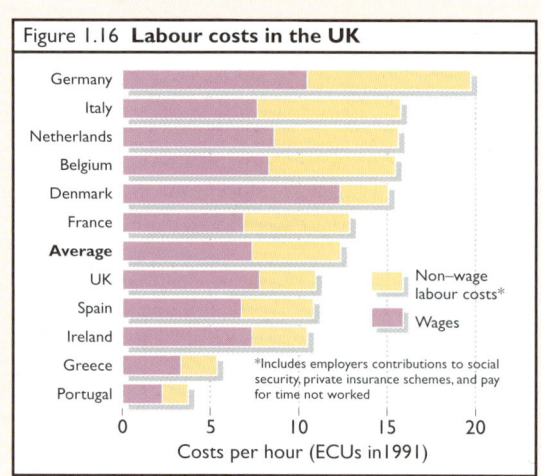

Figure 1.16 Labour costs in the UK

INWARD INVESTMENT

One of the main benefits of the UK's membership of the EU is that it has attracted the lion's share of the foreign multinational corporations (MNCs) which have located in the Union over the last 20 years. The statistics make this clear:

1. The UK has attracted one-third of all inward investment coming in to the EU.

2. Of the £220 billion invested in UK manufacturing over £40 billion has come from overseas.

3. There are now more than 3,500 US companies located in the UK, with 1,000 German companies and around 200 from Japan.

4. 39% of Japanese and 41% of US overseas investment is in the UK.

The UK has used foreign firms to re-industrialise. In contrast to other countries it has welcomed, even courted, foreign investors. Foreign firms can use their British location as a springboard into Europe: their products are considered 'European' as long as they use between 60% and 80% European components in their product. The biggest issue is probably the effect of these 'transplant factories' on the industries of other EU countries. The feeling is that Britain has stabbed its European economic allies in the back in return for jobs, regional regeneration and transfer of technology.

Figure 1.16 shows that the wage and non-wage labour costs of the UK are considerably lower than in other core countries of the EU. The decision by Hoover to rationalise its European operations and close its plant in Dijon to concentrate production in Glasgow was the catalyst for much debate about Britain's stance on the Social Chapter. Britain was branded the 'Taiwan of Europe' by an aggrieved French government, which argued that the UK has sought to achieve an unfair comparative advantage in keeping employment costs artificially low.

GERMAN INVESTMENT IN THE UK

The UK has been the largest recipient of German foreign investment since the late 1980s with £8 billion of investment by German companies between 1990 and 1995. BMW's £920 million takeover of Rover in 1994 was perhaps the most newsworthy and German MNCs in the UK now employ more than 150,000 people. Robert Bösch, the electrical group, located in Cardiff in 1990 to supply components to the UK's car industry. It now employs 2,000 people. Since the late 1980s the electronics firm Siemens has built its workforce from 1,500 to just under 10,000. The benefits are shared.

The UK benefits from transfer of technology, management practices and training. German companies have profited from the UK's lower manufacturing costs and expertise in pharmaceuticals, telecommunications and high technology. They also gain access to the new markets provided by the UK's Japanese and US MNCs.

OUTWARD INVESTMENT

Unlike Germany, the bulk of the UK's overseas investment is outside the EU. The gross total now exceeds £1,400 billion, with an increasing concentration in the emerging economies of South East Asia. Why are British industries not locating in other EU countries? There are several explanations:

1 The UK has a cheaper, less restricted workforce.

2 Some UK industries are locating outside the EU to avoid trade or environmental restrictions.

3 They are attracted to the developing world with its low labour costs and growing markets, especially in South East Asia.

The UK has the reputation of being the most independent of the member states. Its government claims this is because it is the most committed to EU free trade ideals. Others argue that Britain wants the benefits of membership without full responsibility.

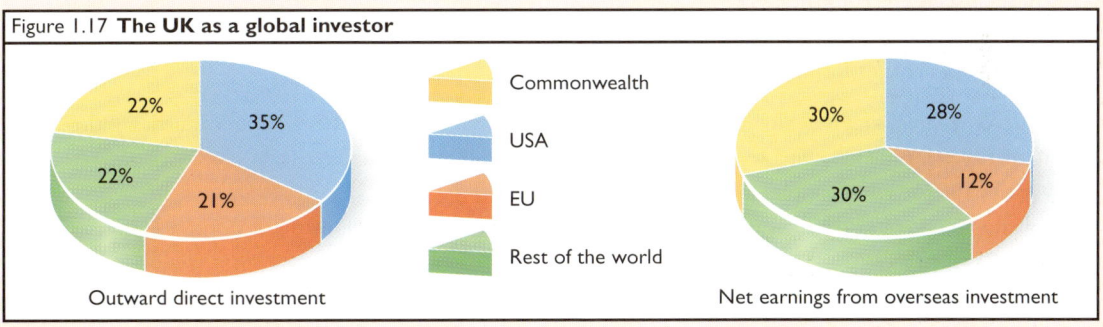

Figure 1.17 **The UK as a global investor**

Outward direct investment: 22% Commonwealth, 35% USA, 21% EU, 22% Rest of the world

Net earnings from overseas investment: 30% Commonwealth, 28% USA, 12% EU, 30% Rest of the world

Figure 1.18 Should we stay...	...or should we go?
Structural Fund money is given to Britain's declining industrial areas and peripheral regions.	An end to budget contributions estimated to reach £2,927 million by 1997.
Freedom to trade within Europe and avoid the EU's common external tariff, currently at about 4%.	A reduction in food prices which are kept high by the tariffs and subsidies of the CAP.
The certainty of trading within the EU allows for forward economic planning.	An increased flexibility with respect to financial and industrial planning.
The attractiveness of the UK to investors from abroad who want to avoid the tariff barriers of the EU.	A chance to maximise trade and profit through outward flows of investment to new markets, e.g. Asia.

Section D Exercises and recommended reading

EXERCISES

Figure 1.19 **Employment structure in EU member states, 1992**

	Employed	Agriculture	Industry	Services	GDP
Belgium	3.8	2.9	30.9	66.2	110
Denmark	2.6	5.2	27.1	67.5	108
France	22.0	5.9	29.5	64.4	113
Germany	36.5	3.7	39.1	57.2	108
Greece	3.7	21.9	25.4	52.8	50
Ireland	1.1	13.7	28.0	58.1	77
Italy	21.0	7.9	33.1	59.0	106
Luxembourg	0.2	3.5	28.9	67.6	131
Netherlands	6.6	4.4	25.6	70.1	103
Portugal	4.5	11.5	32.6	56.0	67
Spain	12.5	10.1	32.7	57.2	78
UK	25.6	2.2	30.1	67.3	99
EUR 12	140.0	5.8	32.6	61.2	100

Employment
Persons in employment (millions)

Agriculture
Employment in agriculture (%)

Industry
Employment in manufacturing (%)

Services
Employment in services (%)

GDP
GDP (base: EUR 12 =100)

All figures for 1992
Source: *Regional Trends, 1995*

1 Define the following terms: *core-periphery, the Single Market, multiplier effect* and *spread effects*. [8]

2 Draw a simple scatter-graph to show the relationship between the percentage employed in manufacturing and the GDP of the twelve EU countries. [4]
   (a) Describe the relationship between these variables. [3]
   (b) How do you account for this? [5]

3 Construct a triangular graph to illustrate the proportions employed in each of the member countries of the European Union (in 1992) for the following sectors: *agriculture, manufacturing* and *services*. [5]
   (a) Describe the pattern of employment that you have shown. [3]
   (b) How do you account for the variations in employment structure throughout the European Union as shown in your graph. [8]
   (c) Evaluate the use of triangular graphs as a tool in geographic enquiries. [5]

4 Choose two contrasting methods of measuring regional inequalities.
   (a) Describe the variations in levels of development they show for Europe. [8]
   (b) Assess the use of these methods as means of illustrating regional disparities. [7]

RECOMMENDED READING

There are a large number of books which focus on the regional geography of Europe. Two useful sources of data are *Competitiveness and Cohesion: Trends in the Regions* by the European Commission (1994) and *Regional Trends* published annually by HMSO. One of the most accessible text books is Barry Brunt's *Western Europe* (1990). More detailed texts include the *Financial Times' Can Europe Compete?* (1994), Pinder's *Western Europe: Challenge and Change* (1990), Dicken's *Global Shift* (1992), and Clout (ed.) *Western Europe* (1989).

CHAPTER 2
THE PRIMARY INDUSTRIES

A **AGRICULTURE**
   The importance of agriculture **20**
   The physical basis **21**
   The human input **22**
   Major European farming types **24**
   The Common Agricultural Policy **28**
   The environmental effects of agriculture **32**

B **THE FISHING INDUSTRY**
   Problems and crisis **36**
   The Common Fisheries Policy **37**

C EXERCISES AND RECOMMENDED READING **40**

Primary industries involve the utilisation of the earth's natural resources. They include all forms of agriculture, forestry, fishing and mining. Primary industries have at times during the mid-1990s appeared to dominate the economic issues of the EU, for example reform of the CAP, the export of live animals to continental Europe and the fish wars between EU members, Canada and Spain. This chapter concentrates on farming and fishing, while mining is examined in Chapter 4 owing to its close links with large-scale manufacturing, especially iron and steel.

Section A explores the nature of farming in the European Union. While physical factors have strongly influenced the initial location of activities, human factors, notably government policies, have shaped their development and success in recent decades. Although farming has declined in importance in terms of employment, it still plays a vital role in Europe's economy and uses up a great deal of the EU's structural and regional funds. The expansion of the EU from twelve to fifteen member countries in 1995 brought new farming types into EU agriculture, notably Alpine and Arctic ones. The possible expansion eastwards into Poland, Hungary and parts of the former Soviet Union is also considered in this chapter, especially in the light of the need for significant land reform. Thus, the issues surrounding European agriculture are as much political as economic, social and environmental.

Section B examines the state of Europe's fishing industry. Whilst not as important geographically or economically as farming, it still plays a vital role in the livelihood of many communities. Many problems and issues exist, notably its sustainability, and the role of the Common Fisheries Policy is analysed as a means of reconciling the interests of conflicting parties.

Section A Agriculture

Agriculture is the use of the land to produce food products for human and animal consumption. It can be classified in a number of ways: intensive or extensive, arable or pastoral, commercial or subsistence and nomadic or sedentary. Europe has a very diverse agricultural base including highly intensive market gardening, extensive pastoral farming in the western fringes and much mixed farming. Although the intensity of production increases towards the core of south east England, north west France, Belgium, the Netherlands and northern Germany, the proportion of workforce employed in agriculture increases towards the periphery and is high especially in Ireland, Portugal and Italy.

THE IMPORTANCE OF AGRICULTURE

Agriculture remains important in Europe for a number of reasons: employment numbers and GDP, food production, trade, land-use and ancillary industries. First, **employment** is important. About nine million farmers and farmworkers were employed in 1990, ranging from nearly two million in Italy to only six thousand in Luxembourg. Nevertheless, all countries have seen a decline in the numbers employed in farming and between 1960 and 1990 the number of farmers in the original six members of the European Union halved. Although the economic importance of agriculture is declining, as measured by contribution to **GDP**, it is still of considerable significance in many countries. On average in Europe it accounts for 2.4% of GDP: it is high in Ireland, Portugal and Greece but is below average in the UK, Germany, Belgium and Luxembourg.

Second, **food production** continues to rise and yields per hectare have also increased. Third, **trade** in agricultural products has escalated: for example, intra-EU trade has risen over twenty-fold, from ECU 4.38 billion in 1968 to ECU 98 billion in 1990, accounting for about 20% of Europe's agricultural produce. Externally, imports exceed exports, by approximately ECU 20 billion in 1990. The main imports are exotic produces, fruit juices and oilseeds.

Fourth, farmland accounts for 57% of the **land-use** in Europe: this varies considerably from 81% in Ireland to 44% in Greece. In general, the flatter countries have more farmland whereas the more mountainous areas, such as Greece and Portugal, have less farmland and relatively more woodland. By contrast, in Belgium and the Netherlands, which are more urbanised and industrialised, there is proportionately less farmland and woodland. Fifth, agriculture provides the basis of many **manufacturing and service industries** such as food processing, agricultural equipment, agricultural inputs (seeds, fertiliser and so on) and a variety of services including finance, vetinary, marketing, and transport.

Figure 2.1 **Small dairy holding, County Kerry, Ireland**

THE PHYSICAL BASIS

Agriculture is influenced by a number of factors, both physical and human. These include climate, soils, drainage, relief, tradition, land ownership, access to capital and markets, demand, technology and government support, among others. In general, physical factors exert a strong influence over what can be grown, while human factors are of increasing importance in terms of what is actually grown and how much.

The diverse nature of farming in Europe can be explained, in part at least, on account of physical environment. Variations in climate, soil, relief and drainage give rise to striking variations in type and intensity of farming, from the market gardening of the Green Heart of the Randstad in the Netherlands to the extensive rough grazing of the Massif Central and the Highlands and Islands of Scotland.

Extremes of **climate** exist between the summer drought of the Mediterranean countries and the cold continental climates of northern Scandinavia. In the former, plant growth is restricted by excessive heat and drought whereas in the latter long winters, frozen soils, low temperatures and snowfall limit growth to a short summer season. In general, Europe's climate varies from a maritime influence in the west, with high all-year rainfall, mild winter temperatures and cool summer temperatures, to a more continental climate in the east, characterised by very hot summers (>25°C) with heavy convection storms, and very cold winters, considerably below freezing, with abundant snowfall. The extremes become more marked as the influence of the Gulf Stream weakens. Agricultural productivity is reduced therefore by excesses of heat in the south, cold in the north, moisture in the west and seasonal extremes in the east.

Geology exerts a considerable influence on agriculture in many parts of Europe. The base-rich soils associated with outcrops of chalk support grasslands, due partly to their high calcium content and partly to their thin soils. By contrast, boulder clay is more prone to waterlogging, but produces fertile soils when drained, as in the case of East Anglia. On the glacial outwash plains, sands and gravels provide acidic podzols poor in agricultural potential: the heathlands of northern Germany and the south of England illustrate this. On the other hand, wind-blown periglacial deposits, *loess*, comprising of silt, have formed the basis of rich *limon* soil in the Paris Basin.

Relief reinforces many of the patterns produced by climate. Indeed, altitude has a direct effect on rainfall and temperature. There is a strong correlation between relief and rainfall, with highland areas experiencing high rates of precipitation. Altitude also lowers temperature (1°C for every 100 metres) and steep gradients have an adverse effect upon the type and level of mechanisation that can be used. Moreover, as mountainous areas are often characterised by poor quality, thin, leached peaty soils agricultural productivity is understandably low. By contrast, in the lower altitude, well-drained river valleys, alluvial silts form the basis of much more productive agriculture.

Figure 2.2 **A sheep farming area, near Grisedale, Lake District**

THE HUMAN INPUT
Land tenure

Tenure refers to the means by which an individual lays claim to a piece of land. There are a variety of types including freehold (ownership), rental (tenancy), share-cropping and state-control. These vary considerably throughout Europe, with ownership being most common in Ireland, Italy and Denmark. By contrast, tenancy is more popular in the UK and Belgium, where up to 75% of the land is worked under tenancy agreements. The leaseholder (owner) may be an individual, a company or even the state: in the Netherlands, the state owns the reclaimed polders and rents them to tenant farmers. Behind the former 'Iron Curtain' the state owned the land and effectively controlled all aspects of food production, from planting to distribution. The key factor is security of tenure: if the farmer has a secure hold on the land for a lengthy period it becomes profitable, and sensible, to make improvements to the farm. By contrast, share-cropping, or *metayage,* as practiced in parts of southern France, is an inefficient form of tenure, whereby the landlord receives a large share of the produce, and there is no incentive for the share-cropper to increase productivity.

Figure 2.3 **A modern dairy parlour**

Figure 2.4 **Field patterns, Otmoor, near Oxford**

Farm size

In general, farm sizes in Europe are small although they vary considerably. Although some extensive sheep farms on the west coast of Ireland and Scotland are large, areas of low productivity, such as the Mezzogiorno and the Massif Central, are characterised by small farms. By contrast, highly productive areas are characterised by large farms, such as East Anglia and the Paris Basin. Since World War II, the number of farms in Europe has decreased by about 50%, and consequently many that remain have become larger. As the ratio of the cost of farm input to the value of return rises small farms have become increasingly **marginalised**. Farms have expanded to achieve economies of scale (savings through increased size) while small inefficient farms have gone out of production. As much as 3% of farms and up to 1% of agricultural land becomes available in any one year. These contrasts are exacerbated as the large farms can afford more inputs, new technologies and thereby achieve economies of scale. By contrast, the small holdings are not in a position to modernise and consequently fall further behind.

Fragmentation, the splitting of an individual farm into many pieces, is also a problem. This occurs as a result of inheritance laws and, increasingly, the expansion of larger farms by the purchase or rental of marginalised farms. Although fragmentation may allow farmers to diversify production, by cultivating a greater variety of environments, it wastes valuable space (in boundaries) and time (in travel).

Access to market

Access to market is an important factor in influencing agricultural productivity. Although it is not as important as suggested in Von Thunen's land-use model (**p. 26**), proximity to market reduces transport costs, increases access to services and facilities, increases land-prices and consequently, productivity and efficiency. This is clearly evident in the core areas of European agriculture. By contrast, peripherality, as in the case of the Mezzogiorno, north west Ireland and Portugal, leads to increased transport costs and reduced access to markets and back-up services, and consequently land-use is less intensive, less profitable and less commercial. Although infrastructural developments have reduced the isolation of some areas, the most peripheral regions continue to remain disadvantaged as a result of their location relative to the main European markets.

Technology and innovations

Exposure to new technology and access to credit facilities can influence the profitability of agriculture. Mechanisation, improved parlour facilities, use of fertilisers and the application of insecticides, pesticides and herbicides led to improvements in yield, quality and reliability. Cereal yields in France and the Netherlands almost doubled between 1970 and 1990, potato yields in Italy and France doubled and there were significant increases in the yields per hectare in oil seed rape and sugar. As intensity of production increased profitability increased and new innovations were introduced, such as new crops or improved strains. The increasing use of **hydroponics** is a good example of a new technology in practise. This is the cultivation of plants, notably high value produce, in water enriched with nutrients, as illustrated in **Figure 2.5** which shows lettuces grown hydroponically in Hertfordshire.

Figure 2.5 **Lettuces being grown hydroponically, Hertfordshire**

MAJOR EUROPEAN FARMING TYPES

Cereal farming

Commercial grain production predominates in flat areas with fertile soils, such as the boulder clay deposits in East Anglia and the loess deposits in the Paris Basin. Summer temperatures are high (≥20°C) favouring growth while winter temperatures are low and help to break up the soil. Rainfall is low (≤700mm) with a slight spring or summer maximum, i.e. during the period of most active plant growth. Farms are characteristically large (≥200ha) with huge rectangular fields favouring heavy machinery. Increasingly farms are run as a business, **agribusiness**, with large financial inputs. **Specialisation** is evident. Access to markets is good and improved transport facilities increases competition between farmers and raises productivity. Cereal farming is highly profitable although it can seriously damage the environment. Highly intensive techniques and the search for greater profits has caused over-production, nitrate pollution, hedgerow removal and soil erosion. As a result, the UK government has classified areas such as the Norfolk Broads, the Brecklands and the Suffolk River Valleys as **environmentally sensitive areas**, in an attempt to preserve important landscapes and encourage less intensive farming.

Dairying

This occurs in four main environments:

1 Hilly coastal locations, e.g. south west Ireland, Normandy.
2 Coastal lowlands, e.g. Denmark, Netherlands.
3 Alpine areas, e.g. Austria.
4 Areas close to large urban markets, e.g. the Low Countries.

The similarities and contrasts are shown by **Denmark** and **south west Ireland**. In Denmark, intensive dairying predominates due to close access to large urban markets. It is a high value form of agriculture, based on small (<30ha) family farms, organised and controlled by co-operatives. Less than 10% of Denmark is pasture land: animals are fed mostly on fodder crops, especially barley. By contrast, dairying in south west Ireland is far less intensive, and up to 80% of the land is pasture. This is mostly due to climate: rainfall is high (≥1,500mm) with mild winters (≥8°C) and cool summers (≤18°C). This favours all-year grass growth and prohibits large-scale cereal cultivation. Like Denmark, farm size is generally small (≤20ha), family owned and the cooperative movement has played a vital role in collecting, processing and marketing dairy produce.

Sheep rearing

This is associated with upland areas which are too steep, cold and wet for arable or dairying. It is also associated with chalk downlands where the thin alkaline soil favoured grass growth and prohibited the use of machinery although improvements in technology have now overcome this (p. 34). A combination of high rainfall (≥1,500mm), low temperatures (summer average ≤15°C, winter average ≤3°C), steep slopes and thin, infertile, leached soils greatly limit agricultural options. As plant productivity decreases the need for more acreage per livestock increases. **Crofting** in the Highlands region of Scotland is an extreme form of extensive sheep rearing. Farm sizes are very large (frequently ≥400ha), owing to impoverished grazing. Returns are low and without national and EU aid many more sheep farmers and crofters would cease farming. Most sheep farmers in the British Isles are eligible for **subsidies** for each breeding ewe, receive EU aid for Less Favoured Areas status and many have diversified into non-agricultural activities such as tourism, educational visits and part-time work.

Beef rearing

This is concentrated in less fertile upland areas unable to accommodate dairying such as the **Mezzogiorno**, the **Highlands of Scotland** and the **Massif Central**. Owing to high rainfall, steep slopes and low temperatures soils and pasture are poor. Hence farms needs to be large to provide sufficient grazing per cow. Beef rearing in the Massif Central epitomises this type of faming. A mixture of steep, dissected relief, poor soils, low winter temperatures and high summer temperatures limits potential. **Small fragmented farms** predominate; many farmers are elderly and conservative and the region is relatively isolated from the main markets. Some improvements to the area's economy has been evident a result of planning initiatives. In 1962 SOMIVAL was established with the aim revitalising the region. Some successes have been observed: there are now fewer but larger farms; increased specialisation on Charolais and Limousin cattle better adapted to the local environment; improvements in grazing quality via fertilisers and improved grass species; and off-farm changes, such as increases in the income derived from tourism and forestry and more general improvements in the rural infrastructure.

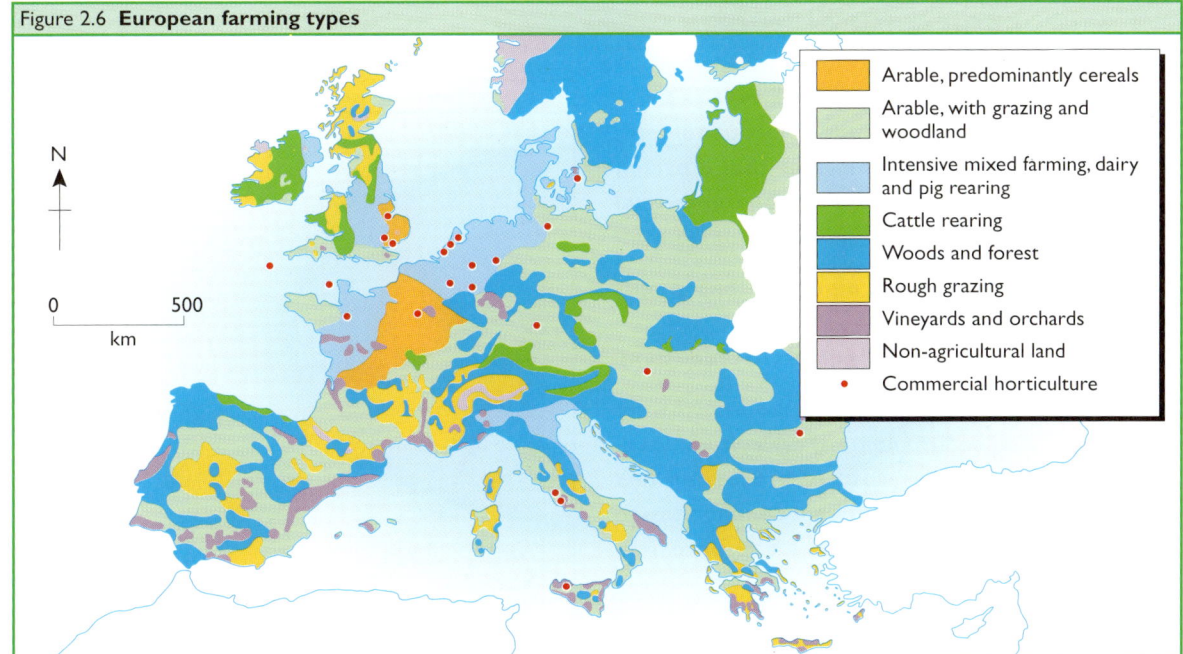

Figure 2.6 European farming types

Mediterranean agriculture

This has traditionally been characterised by **extensive** production of cereals, olive trees and vineyards. Farming organisation was characteristically poor with over 38% of farmland in Europe but over 50% of all farmers. Productivity is limited by a number of factors:

1. Rugged relief: 80% of Italy and Iberia is mountainous.
2. Harsh climate: low, unreliable rainfall with a summer drought.
3. Poor soils and highly denuded slopes.
4. Isolated location to the major markets.
5. Underdeveloped transport facilities.
6. Limited industrialisation.
7. Poor demographic structure: high proportion of elderly and undereducated.
8. Inefficient land tenure: a mixture of latifundia—large estates frequently with absentee landlords—and minifundia—tiny fragmented subsistence holdings.

However, change has occurred, notably along the coastal belt. A combination of national and EU aid, agrarian reform, such as changing land tenure and irrigation, and improved transport infrastructure have overcome the limits of the coastal areas. In the irrigated, lowland areas such as the areas surrounding Naples and Bari, farming has intensified and specialised. Tree crops such as olives, vines and citrus and vegetables have increased in importance, and cereals such as maize and wheat dominate the cropping pattern. By contrast, upland areas in the Mezzogiorno, are still characterised by extensive farming and low productivity. Sheep, goat and beef cattle are the main forms of livestock. Farming is still largely subsistence with commercial farming limited to small pockets.

Market gardening

This is a form of highly **capital intensive** farming based on very small farms (≤10ha) producing high quality fruit, vegetables and flowers. It is found in areas close to large urban markets, such as the Vale of Evesham in relation to Birmingham, and where physical factors favour early growth, e.g. south west facing slopes on the edge of the Mendips, Somerset. Physical factors enable high productivity but the combination of human factors explains its highly developed nature. These include high demand for products, high cost of land and limited land availability, close proximity to a wealthy urban market, well-developed transport network and government policy at national and international level. Market gardening in the Randstad is considered in detail on **pages 106–7**.

INSET 2.1 AGRICULTURAL MODELS

VON THUNEN'S LAND-USE MODEL

This is based on the concept of locational rent, the maximum amount of income that a person can derive from a piece of land for commercial farming. In his model, Von Thunen assumed that:

1. All farmers wished to maximise their profits ('rational man').
2. There were no physical barriers to production.

Thus, access to market and transport costs were the overriding determinants of agricultural land-use. As market prices, costs of production and transport costs vary for different commodities so too will their profitability vary with distance from the market:

1. High value, perishable goods dominate the agricultural landscape close to the market.
2. Low value, bulky products dominate further away from the market.
3. High land-use intensities are found close to the main agricultural markets.
4. Lower land-use intensities exist further away from the markets.

This was true even when Von Thunen introduced a navigable river and a secondary market into his model: it distorted the original pattern, but the principles and general features remained the same.

Although improvements in transport and storage have made the model largely redundant in developed countries it is still a useful starting point as it allows us to concentrate on the economic side of agriculture. It retains a certain degree of validity, especially in areas with underdeveloped transport facilities, as in Europe's periphery, although here the problem is compounded by adverse physical conditions. On a small-scale, such as an individual farm, his principles can be seen: high land-use intensity close to the farm and lower intensities further away from the farm buildings.

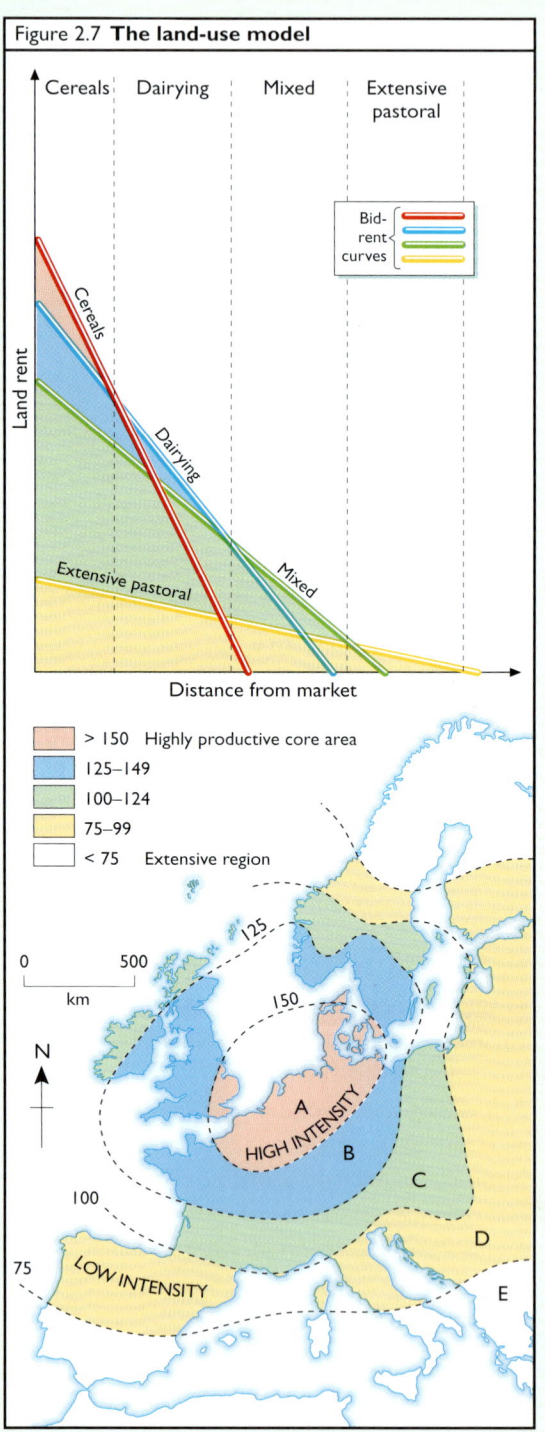

Figure 2.7 The land-use model

SINCLAIR'S LAND-USE MODEL

By contrast, Sinclair found that land-use close to urban markets was dominated by low levels of agricultural intensity and that the value of agricultural land, for agriculture, increased with distance form the market. The reason for this is clear: close to the urban areas there is much speculative development. The land is more valuable for housing, commercial or transport developments rather than agricultural uses. Therefore, there is little investment in agriculture and the land remains under-utilised. There may also be a problem of theft or trampling of crops by urban residents. However, there is a pronounced distance decay and beyond a certain distance neither speculative development, trampling nor theft exert significant effect upon the value of land.

Figure 2.8 Farming on the urban fringe
The zone on the edge of the city is referred to as the 'urban fringe' or peri-urban area. Agricultural characteristics of peri-urban areas include:
A high proportion of the total land area is farmland.
Much of the land is used for horticulture, dairying, poultry, pigs or mixed farming.
Very little land is left fallow.
Labour/capital inputs are very high; outputs are high.
Land-use and labour is intensive.
Cost of land is high and increasing rapidly.
Most farms are small and many are fragmented.
Tenancy arrangements and part-time farming are common, especially closer to the urban area.
Close proximity to urban populations has led to a high proportion of 'farm gate' sales and 'pick your own' farms.
Output may be reduced due to vandalism, green belt restrictions, speculative developments and environmental regulations on smell, farm waste and so on.
Other drawbacks include construction of ring roads, industrial pollution and high wages for casual labour.

THE BEHAVIOURAL MATRIX

The models above suggest that farmers will behave in a particular fashion, i.e. they will do whatever will maximise their profit. However, not all farmers necessarily wish to do so and even if they did they may not have the ability or the information to do so. As a reaction against 'rational man' and scientific predictability, 'behavioural man' was introduced, a person with imperfect knowledge and ability, behaving in a subjective manner. The behavioural matrix simplifies behavioural man to two axes, ability to use information and quality and quantity of information. As the ability and information increase then the farmer should make the 'right' decisions whereas a farmer with limited information and ability would be expected to make the wrong decisions. This is, of course, almost as predictive as the scientific method, and a great simplification of behavioural man. However, it does have some merit: an MNC with huge financial resources and specialist workers would expect high quality information and have the ability to use it. By contrast, a peasant farmer with little schooling, training or financial resource would not be expected to make the right decisions, although he may do so by chance.

HAGERSTRAND'S DIFFUSION CURVE

Hagerstrand was interested in the diffusion of innovations and incorporated a chance element in his model. He showed that the introduction of a new technique depended upon a number of factors including information regarding innovations, financial security, psychological make-up of the adopter, and physical proximity to other adopters. Initially very few people adopt an innovation. As information becomes more widespread, and often the cost is reduced, increasingly more people adopt. However, there will always be some who are adverse to change and will take a long time, if at all, to accept the new technique.

THE COMMON AGRICULTURAL POLICY

There were three basic principles behind the formulation of the CAP:

1. The **unity of the market**: a single agricultural market within which goods move freely.
2. **EU preference**: products grown within the EU should be purchased rather than those originating outside the EU.
3. **Financial solidarity**: EU financing of CAP.

In 1957 the **Treaty of Rome** was signed establishing the European Economic Community (EEC). Article 39 set out the main priorities for agriculture:

- To increase agricultural productivity and self-sufficiency.
- To ensure a fair standard of living for farmers.
- To stabilise markets.
- To ensure that food was available to consumers at a fair price.

Between 1958 and 1968 these aims were implemented: a single market existed in agriculture from 1962 and a common set of market rules and prices were introduced by 1968.

At the centre of the CAP was the system of **guaranteed prices** for unlimited production. This encouraged farmers to maximise their production as it provided a **guaranteed market**. By 1973 the EU was practically self-sufficient in cereals, beef, dairy products, poultry and vegetables. In order to maintain the principle of EU preference European products had to remain cheaper than imports. Imports were therefore subjected to import duties or levies and exports subsidies were introduced to make community products more competitive on the world market. Despite its element of EU preference, the CAP has not prevented farm imports into the EU from growing and the EU imports more food than it exports, giving it a trade deficit of some ECU 20 billion.

CAP led to intensification, concentration and specialisation. **Intensification** is the rising level of inputs and outputs from the land as farmers sought to maintain or increase their standards of living (or margins of profitability). The inputs included fertilisers, animal feed, fuel and machinery. The increased levels of outputs were typified by beef and butter 'mountains' and 'wine lakes'. **Concentration** is the process whereby production of particular products has become confined to particular areas, regions or farms. **Specialisation** is related to concentration and refers to the proportion of total output of a farm, region or country accounted for by a particular product. For example, wheat has become more concentrated in France and the UK as farmers have specialised in it.

Figure 2.9 **The need for the Common Agricultural Policy, and some of its effects**

AGRICULTURE 29

Figure 2.10 CAP: The effect on West Midgeley Farm

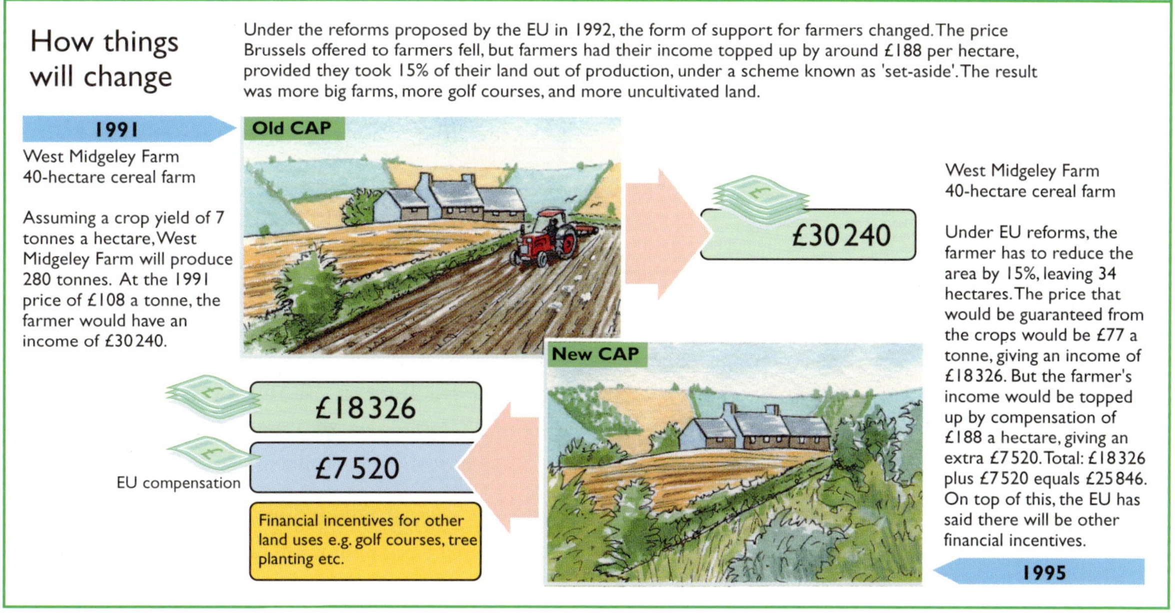

How things will change

Under the reforms proposed by the EU in 1992, the form of support for farmers changed. The price Brussels offered to farmers fell, but farmers had their income topped up by around £188 per hectare, provided they took 15% of their land out of production, under a scheme known as 'set-aside'. The result was more big farms, more golf courses, and more uncultivated land.

1991

West Midgeley Farm
40-hectare cereal farm

Assuming a crop yield of 7 tonnes a hectare, West Midgeley Farm will produce 280 tonnes. At the 1991 price of £108 a tonne, the farmer would have an income of £30 240.

Old CAP → £30 240

New CAP ← £18 326

EU compensation ← £7 520

Financial incentives for other land uses e.g. golf courses, tree planting etc.

West Midgeley Farm
40-hectare cereal farm

Under EU reforms, the farmer has to reduce the area by 15%, leaving 34 hectares. The price that would be guaranteed from the crops would be £77 a tonne, giving an income of £18 326. But the farmer's income would be topped up by compensation of £188 a hectare, giving an extra £7 520. Total: £18 326 plus £7 520 equals £25 846. On top of this, the EU has said there will be other financial incentives.

1995

The need for reform

In recent years there has been a reform of the Common Agricultural Policy because price guarantees and intervention storage created **surpluses** in cereals, beef, wine and milk. Indeed, by the 1990s the EU was overproducing cereals by 20% while demand had dropped. Whereas production increased from 1973 at an average rate of 2%, demand increased at only 0.5%. The average EU household spent 21% of their household income on food compared with 28% in 1970. In some sectors, **technological and scientific** improvements improved yields, further increasing surpluses. Consequently, a larger proportion of EU funding was used to store and sell-off surpluses at subsidised prices on the world market. Moreover, the EU has also lost some of its traditional export markets in the former Soviet Union and parts of the Middle East, thereby reducing **demand**.

The first modifications came in 1979 with other major changes introduced between 1984 and 1988. In 1979 a **co-responsibility levy** was introduced on dairy farmers to help meet the cost of storing the surplus and selling it off cheaply on the world market. This levy proved insufficient and in 1984 a system of **quotas** was introduced. These were reinforced by 1988 '**stabilisers**', mechanisms to control EU spending on each product. These have only been partially successful, and surpluses have continued to rise.

Spending on CAP

Although the cost of the CAP rose from ECU 26 billion in 1988 to ECU 36 billion in 1992, farm incomes suffered. Moreover, as price support was proportional to production, 80% of EU spending was directed towards 20% of the farmers, the bigger and more efficient ones. **Small family-farms** became increasingly **marginalised** under CAP as they were unable to benefit from new technologies, intensive production methods and could not achieve economies of scale. Consequently, many farmers left the land.

However, in relative terms, spending on the CAP has been reduced: this reflects an increased amount of spending on other sectors, notably regional and social development. Many of the former responsibilities of the CAP have been taken over by the EU's regional and social funding. Since 1988 the Guidance Section of the European Agricultural Guidance and Guarantee Fund (EAGGF), the European Regional Development Fund (ERDF) and the European Social Fund (ESF) have diversified and promoted rural areas. For example, the EAGGF provides support for farmers and funds environmental protection schemes, the ERDF concentrates on infrastructural development and the ESF provides training and job creation programmes.

The CAP after 1992

By 1992 the CAP had achieved some of its aims such as the regular **availability of supplies, stabilisation of markets** and **reasonable prices**. However, there was a still a need for reform since the cost of subsidies were exceedingly high, representing an average subsidy of approximately £257 per person in Europe (1993), £10,000 for every agricultural worker and £653 for every agricultural acre.

The most important changes to the CAP were introduced in 1992. Five objectives were identified:

1. To increase Europe's competitive agricultural base.
2. To match production with demand.
3. To support farm incomes.
4. To stop the drift out of agriculture.
5. To protect and develop the potential of the natural environment.

To achieve this, a variety of changes were introduced:

- Reduction of price support where surpluses exist.
- Increased quotas on milk, wine, cereals and olive oil.
- Guaranteed maximum quantities.
- Concentration on quality rather than quantity.
- Alternative rural land-uses.
- Extensification of land-use.
- Income support to farmers in less favoured areas.
- Early-retirement schemes.
- Training and assistance for young farmers.

The key elements are **price cuts** and the **withdrawal of land** from production. For example, milk quotas were reduced by 2% and the price for cereals and beef dropped 29% and 19% respectively, between 1993 and 1996. Consequently, there is more incentive to diversify agriculture and to make it less intensive. However, it will take considerable time to see whether these prove successful. In theory the reforms should make farmers more competitive and farmers should gain access to more markets. Thus, it is possible that even with a reduction in price support and set-aside, cereal harvests and beef yields will increase, especially from the aggressive and innovative farmers. On the other hand, there has been an increase in income support: for example, in the early 1990s over 700,000 farmers received the suckler cow premium and over 600,000 the ewe premium. This may be keeping some marginal farmers in production.

However, there are a number of key questions:

1. Will farmers become more dependent on hand-outs?
2. Will reform lead to a drop in farm income?
3. Will there be any decrease in the CAP budget?
4. What affect will reform have on the countryside?

There has also been reform of the rural development programme. Structural funds, worth almost ECU 18 billion in 1992, have been concentrated in two main areas:

1. **Objective 1 areas**: backward areas where all development schemes are supported.

2. **Objective 5b areas**: backward rural areas where only rural development areas are subsidised by the structural and guidance funds in particular.

Moreover, special concessions are made to farmers in mountainous areas and in less favoured areas with adverse soil and weather conditions, in all about 50% of farmland in Europe. Many of the development programmes have been concentrated on the peripheral parts of Europe, notably Iberia, Ireland and Greece. and over three hundred projects, aimed at improving transport, communications, water supplies and energy distribution, have been supported, alongside farm modernisation, training and the acquisition of new skills.

Figure 2.11 **Oil seed rape, Oxfordshire**

Figure 2.12 **The reform of the CAP and global food supplies**

The food is running out

The worst food crisis since 1974 has left the world with only 53 days supply of grain. The world is seriously short of food and up to 35,000 children die from hunger related diseases every day. Worldwide grain stocks are well below the FAO's minimum necessary to safeguard world food security. In 1987 there were over 100 days' worth of food supplies, but in 1995 there were just over 50, and by the end of 1996 it will be below 50 days'.

World food production has lagged behind food consumption since 1993. The drought of 1995 has led to the lowest harvest of food/head since the mid-1970s. The blistering summer of 1995, the hottest recorded in many parts of the northern hemisphere, destroyed millions of crops. The drought in Spain entered its fourth year and wheat yields slumped to less than half of their 1994 levels.

The European food mountain is fast being eroded. Its grain mountain has fallen from 33 million tons in 1993 to just 5.5 million tons in 1995. The only significant surplus is the wine lake, at 120 million litres. In Britain, in 1994 there were over 1 million tons of grain stored; in 1995 150 tons of barley. Airport hangars in 36 locations around the UK, such as those at Tangmere near Bognor Regis, have stored the UK's contribution to the EU food mountain. Now they are all but empty. Europe is not producing enough food, partly as a result of the 1992 CAP reforms which increased the amount of set-aside. Set-aside land is now being put back into production. Plans to reduce set-aside to 10% of land, rather than 15% in 1995, are an attempt to increase food production.

The world food crisis has slashed stocks and set prices rocketing and world food prices have risen above the politically inflated European ones. For the first time in over twenty years the EU is set to tax European food exports rather than subsidise them, to prevent them from being bought up quickly by the rest of the world.

Rising prices have a disproportionate effect on poor countries who need to find an extra £2 billion just to be able to pay for the same amount of food next year as they have had this year. Food aid is also likely to suffer as producer countries have little extra to give away. Famine and global distress are ominously close.

The way forward?

European farming in the twenty-first century is likely to be very different from that of today. How effective the reform of the CAP will be is open to question. It seems inevitable that more reforms will occur although many farmers are against any further change. The 1992 changes will not resolve the issues in European agriculture and it will take a number of years even to implement the proposals. Moreover, the success of these policies depend not only on conditions within Europe but at an international scale too. The accession of Austria, Finland, and Sweden to the EU in 1995 has had a neutral effect financially whereas the proposed absorption of central and eastern European countries could have an increasingly draining effect. Many of these agricultural economies are now in disarray as a result of the loss of traditional Soviet markets and the collapse of the state distribution system. Until the Single Market, many peripheral countries were protected from competitors, but are now subject to cheap imports from the agribusinesses of the core (**p. 149**). Moreover, the continuing concern with environmental issues, especially in northern Europe, is another force demanding reform of the CAP and European agriculture.

Agriculture is a highly complex system and there are a number of apparent paradoxes, for example the increasing development of technology to increase yields alongside quotas, stabilisers and set-aside to reduce production. It is likely that farming will be intensified in the more favoured areas, the core, and land will continue to come out of production in the less favoured areas. Europe's decreasing number of farmers are using less land and adopting more technology. At the same time they are stabilising and diversifying production in an increasingly competitive fashion.

THE ENVIRONMENTAL EFFECTS OF AGRICULTURE

Agriculture has had a variety of effects on the environment for a considerable length of time. Early neolithic farmers were involved in burning forests to create pasturelands, Anglo-Saxons had the use of iron to plough heavy clay soils and the late Mediaeval enclosures turned an open field system into a patchwork of fields and dispersed settlement. Modern farming continues to wield an impact on the environment. This includes the use of fire, hedgerow removal, the application of chemical fertilisers, the disposal of farm waste and the use of heavy machinery. Two examples from the UK are examined here in detail: soil erosion and nitrate pollution. The British government has responded to the environmental impact with a number of schemes. Along with the expansion of set-aside and Environmentally Sensitive Areas (**ESAs**) a number of measures addressing specific issues have been introduced:

1. **Nitrate Sensitive Areas** to protect groundwater areas.
2. **Habitat schemes** to improve/create wildlife habitats.
3. **Organic aid schemes** to encourage farmers to convert to organic production methods.
4. **Countryside access scheme** to grant new opportunities for public access to set-aside land and suitable farmland in ESAs.

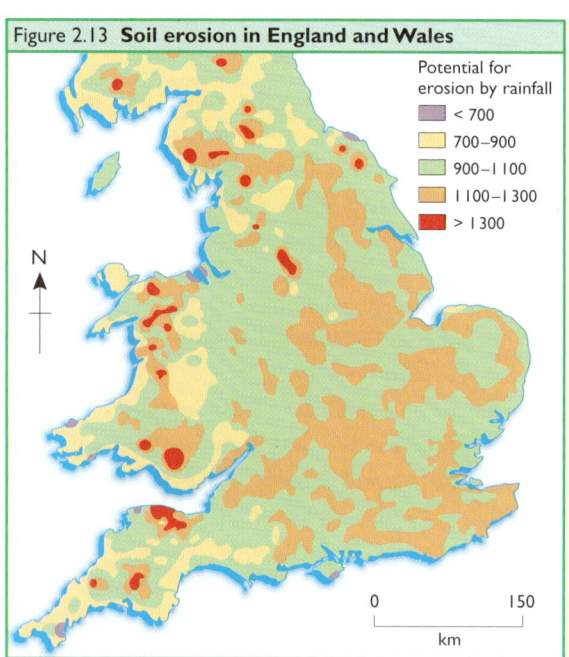

Figure 2.13 **Soil erosion in England and Wales**

Figure 2.14 **Wind erosion in Oxfordshire**

Soil erosion

This is widespread in Europe on **sandy**, **loamy** and **peaty** soils which are vulnerable to erosion by wind and water. Since 1945, the potential for **overland run off** and soil erosion has increased as pasture has been converted to arable land and more winter crops are sown. For example, on the South Downs, traditional sheep pastures have been replaced by arable fields on slopes as steep as 20°. The switch to winter crops in the mid-1970s was accompanied by hedgerow and bank removal and **field enlargement**. The use of heavier, more **powerful machinery** not only compacts the soil but creates 'tramlines' for overland run off to follow. The use of fire to burn stubble (which was eventually banned in the early 1990s) briefly enriched the soil but removed organic content from the soil. This organic material bound the soil and helped it resist erosion. The optimum growth of winter crops requires light, fine soils and the selective use of herbicides. However, fine soils are easily eroded whereas coarser soils with stubble can resist most storms. The consequences are striking: soil losses of up to 250t/ha have been recorded and gullies several metres deep have been initiated in storms. Elsewhere in the UK soil erosion is a familiar problem. In the Midlands it is associated with potato and sugar beet fields in the summer months after the harvest. Soil erosion is also a major problem in South Limburg (Netherlands), central Belgium and the Paris Basin. The potential for soil erosion largely depends upon whether there is a return to grass or whether arable crops continue to expand—and this depends to a large extent on the CAP.

The nitrate issue

Nitrogen is a key component for plant growth and so farmers are keen to apply nitrogen fertilisers. Moreover, national and European policies promoting agricultural self-sufficiency and the manufacturers of nitrate fertilisers are also keen to see an increase in their use. Their use in the UK rose from just 200,000 tonnes in 1945 to a peak of 1.6 million tonnes in the late 1980s. However, there are serious ecological, economic and health effects and recent legislation has curtailed their use.

Eutrophication, or nutrient enrichment, of water bodies has led to algal blooms, oxygen starvation and a decline in species diversity. This is most evident in poorly circulating waters, especially ponds and ditches. While there is a strong body of evidence to link increased eutrophication with increased use of nitrogen fertilisers, some scientists argue that increased phosphates from farm sewage are the cause.

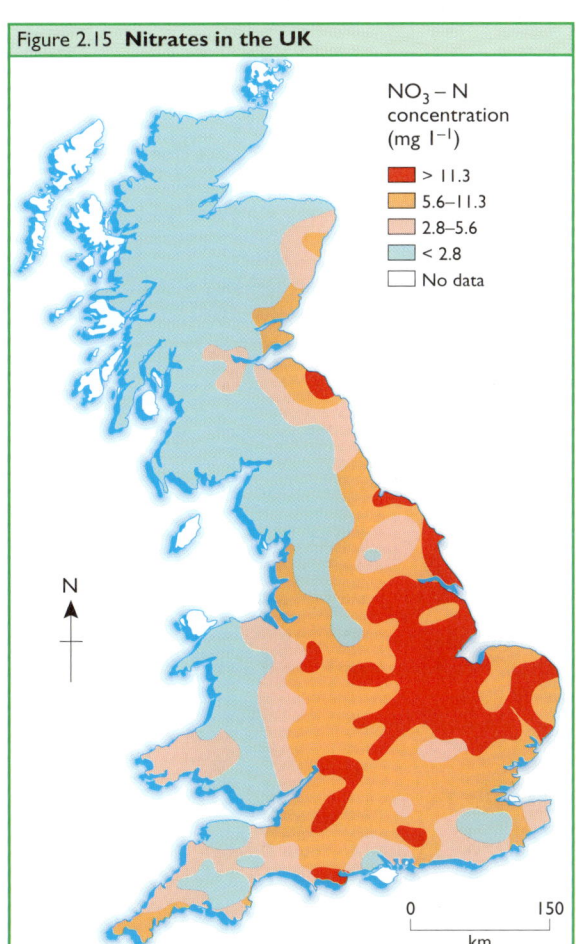

Figure 2.15 **Nitrates in the UK**

$NO_3 - N$ concentration ($mg\ l^{-1}$)
- \> 11.3
- 5.6–11.3
- 2.8–5.6
- < 2.8
- No data

The concern for health relates to increased rates of **stomach cancer**, caused by nitrates in the digestive tract, and blue baby syndrome, **methaemoglobinaemia**, caused by oxygen starvation in the bloodstream. However, critics argue that the case against nitrates is not clear: stomach cancer could be caused by a variety of factors and the number of cases of blue baby syndrome is statistically small.

Of more general concern is the amount of nitrates in tap water. The pattern of nitrates in rivers and groundwater shows marked regional and temporal characteristics. In the UK, it is concentrated towards the arable areas of the east, and concentrations are increasing. In England and Wales over 35% of the population derive their water from the aquifers of lowland England and over 5 million people live in areas where there is too much nitrate in the water. The problem is that nitrates applied on the surface slowly make their way down to the groundwater zone, and this may take up to 40 years. Thus increasing levels of nitrate in drinking water will continue to be a problem well into the twenty-first century. The annual cost of cleaning nitrate-rich groundwater is estimated at between £50 million and £300 million.

Since the late 1980s the problem has been tackled in a number of ways:

1. Changing land-use; less arable land, either due to set-aside, wood lots or pastoral farming.
2. Changing inputs; extensification of agriculture.
3. Avoiding the use of nitrogen fertilisers between mid-September and mid-February when rainfall is higher.
4. Giving preference to winter crops.
5. Sowing cover crops early.
6. Applying fertilisers in early spring when plants need it most.
7. Avoiding riparian (riverside) fields.
8. Not applying if the weather forecast is heavy rain.
9. Using less nitrogen fertiliser if the previous year was dry.

During 1994, the UK government published proposals for implementing the EU nitrate directive, aimed at reducing existing pollution and preventing further pollution. A total of 72 Nitrate Vulnerable Zones (NVZs) were identified in England and Wales, covering 650,000ha and four in Scotland covering 70,000 hectares. Measures to reduce pollution will take effect between 1996 and 1999.

Protecting environments: set-aside and environmentally sensitive areas (ESAs)

The set-aside scheme was introduced on a voluntary basis in 1988 allowing farmers to take up to 20% of their land out of production and to receive up to £200 for each hectare set aside. The land could be left fallow, converted to woodland or used for non-agricultural production. Reform of the CAP in 1992 reduced the amount of set-aside to a maximum of 15% and further reform in 1994 reduced rotational set-aside to 12% and flexible set-aside to 15%. While many farmers took advantage of set-aside, many intensified production on the other land and made their least favourable land the set-aside! Between 1992 and 1993 the total area in England and Wales under cereals decreased by 400,000ha and there was a similar increase in the amount of set-aside.

In 1985 the EU agreed to provide farmers with the means to farm **environmentally sensitive areas** in traditional ways which would preserve important biological and heritage landscapes. Less intensive, organic methods were favoured with increased amounts of fallow. By 1994, 10,500 farmers had signed or applied for ESA agreements, and payments during 1994-5 totalled about £25 million.

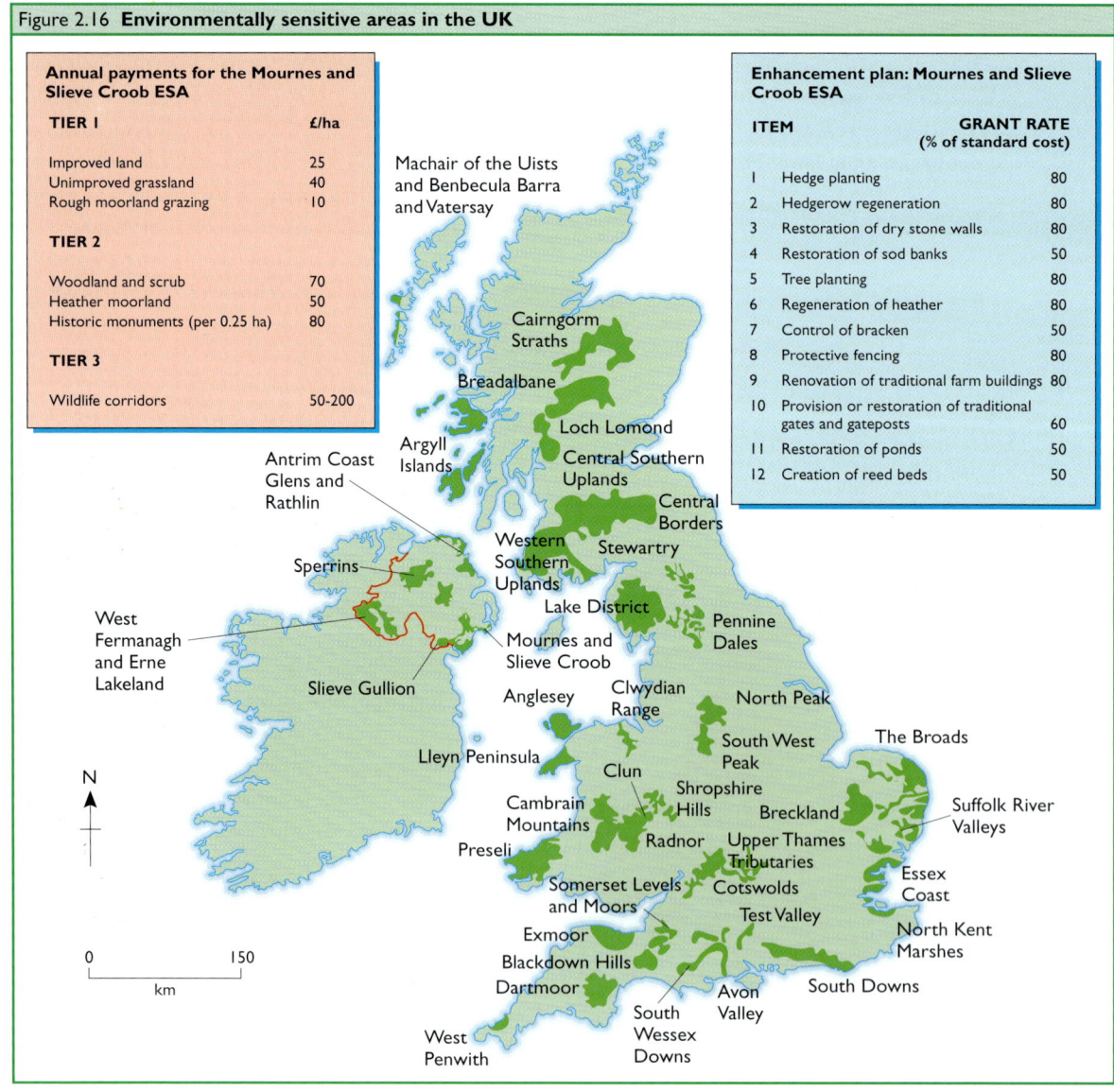

Figure 2.16 Environmentally sensitive areas in the UK

Annual payments for the Mournes and Slieve Croob ESA

TIER 1	£/ha
Improved land	25
Unimproved grassland	40
Rough moorland grazing	10

TIER 2	
Woodland and scrub	70
Heather moorland	50
Historic monuments (per 0.25 ha)	80

TIER 3	
Wildlife corridors	50-200

Enhancement plan: Mournes and Slieve Croob ESA

ITEM		GRANT RATE (% of standard cost)
1	Hedge planting	80
2	Hedgerow regeneration	80
3	Restoration of dry stone walls	80
4	Restoration of sod banks	50
5	Tree planting	80
6	Regeneration of heather	80
7	Control of bracken	50
8	Protective fencing	80
9	Renovation of traditional farm buildings	80
10	Provision or restoration of traditional gates and gateposts	60
11	Restoration of ponds	50
12	Creation of reed beds	50

Figure 2.17 **Land-use options in the Pennines**

Today's landscape
This typical Dales scene supports a community earning a living from farming or tourism. It is the product of an agricultural system that was, until recently, supported by subsidies intended to increase food production and maintain a healthy rural economy. Some meadows are cut for hay, while others are used for silage production. Some walls and field barns remain in good order, but many derelict ones are replaced by fences or modern sheds. Broad-leaved woodlands are being damaged by stock and some heather moorland is deteriorating due to overgrazing.

Without subsidies; extensification
In this landscape future farming subsidies have been taken away, leaving upland farmers to compete with better farms in the lowlands. In this situation, many owner-farmers would sell-up, while tenanted farms would be taken back into estates and their buildings used for alternative purposes. The few remaining farmers would keep smaller flocks on improved land, but outlying meadows and pastures would be abandoned. With no money to maintain field barns, walls, and woodlands, they would decay and become derelict. To survive some farmers would turn to farm-based tourism or forestry.

Intensification
This landscape, geared to food production, is supported by no public money. With access to the latest technology and breeding techniques, larger farmers could still make a profit by buying out the smaller farmers to create big livestock ranches. The flowery meadows would be intensified and the grass taken mainly for silage. Large sheds and wire fences would replace walls and field barns. Broad-leaved woodlands would die as a result of stock damage and extensive conifer woodlands would be planted. Some heather moorland would convert to grassland due to overgrazing.

Diversification
If farming subsidies are withdrawn the traditional estate may become the dominant economic unit, with game shooting and leisure providing more income than farming. Some livestock rearing would still be carried out by tenant farmers, although many owner-farmers would have sold up. Tenants would be encouraged to improve habitats, such as moorland, broad-leaved woodlands, and meadows, for game and wildlife. Some barns would be converted to new uses and estate owners would diversify into local industries and tourist activities, like nature trails or riding.

Section B The fishing industry

The European Union is the world's largest market for fish products and the third major sea fishing power behind Japan and China. Overall, the contribution of fishing to the economy of the member states may appear to be limited. In most countries it represents less than 1% of GDP and provides comparatively few jobs (Figure 2.18). Of far greater importance is its vital contribution to certain coastal communities where it provides jobs and incomes directly and also in back-up industries, like boatyards, processing, packaging and equipment suppliers. There are estimated to be around 30,000 full- and part-time fishermen in the EU and every job at sea creates a further four to five on shore. Figure 2.18 shows the importance of fishing to the peripheral states, Spain, Portugal and Greece, and to the Scandinavian members.

PROBLEMS AND CRISIS

Increased public demand, pressure on fishermen to cover rising investment costs and the development of new technology have created a crisis in the European industry. A vessel can now trawl four nets rather than one; factory ships can freeze and process hundreds of tonnes of fish before returning to port; satellites and sonar identify shoals and help fishermen lay nets. The result is that fishermen try to catch as many fish as possible before someone else does (**Inset 2.2**). At the same time this new technology can be hugely wasteful with respect to money and natural resources. It may also cause environmental problems which can directly affect fish numbers in two ways:

1. **'By-catch'**: large numbers of fish may be trapped in nets but are too small or the wrong species and so are thrown back dead into the sea with disastrous effects on breeding stocks (discards can account for more than half a catch).

2. **Seabed damage**: the use of beam trawlers and other fishing gear with heavy chains can damage fish spawning and feeding areas on the seabed.

These problems have a direct effect on the **sustainability** of existing fish stocks. The European fishing industry also has an **overcapacity** of vessels which leads to overfishing which in turn depletes stocks even further. In response to this depletion in fish stocks and problems arising from increasing competition outside the EU, the European Commission established the Common Fisheries Policy in 1983.

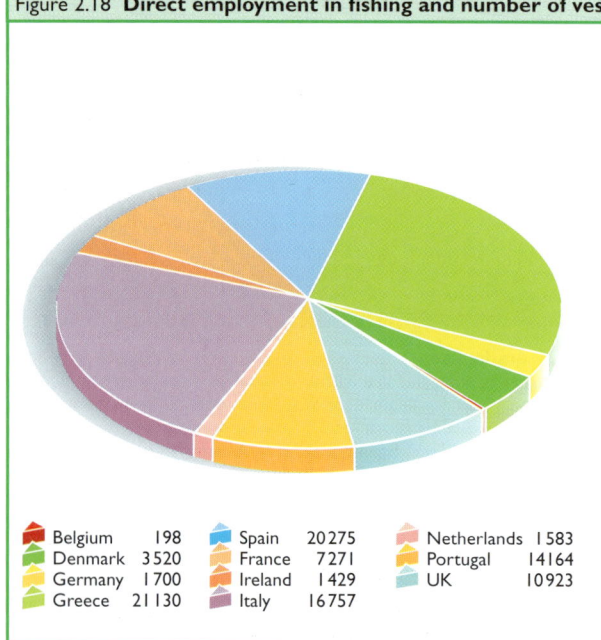

Figure 2.18 Direct employment in fishing and number of vessels

Belgium	198
Denmark	3520
Germany	1700
Greece	21130
Spain	20275
France	7271
Ireland	1429
Italy	16757
Netherlands	1583
Portugal	14164
UK	10923

No. of fishermen	1989	1990	1991
Belgium	908	845	818
Denmark	7317	7317	6886
Germany	1898	4812	4291
Greece	40164	40164	40164
Spain	88199	87351	84838
France	34097	32622	30971
Ireland	7900	7900	4949
Italy	49766	49766	49766
Netherlands	4000	3502	3932
Portugal	40996	40601	38507
UK	22217	24230	24230
EUR 12	**297462**	**299110**	**289352**

INSET 2.2 THE 'TRAGEDY OF THE COMMONS'

Renewable resources, such as fish, need not be depleted provided that the rate of use does not exceed maximum sustainable yield. In other words, that the rate of use is within the limit of natural replacement and regeneration. If resources become over-exploited, then depletion and degradation will lead to scarcity. If more than one nation is exploiting a resource, which is clearly the case in the fishing industry, resource degradation is often the result. Garrett Hardin (1968) has suggested a metaphor, the 'Tragedy of the Commons', to explain this tendency:

In medieval England a number of farmers are grazing cattle on common land. The carrying capacity of the land has been met; more animals would lead to overgrazing, but fewer would mean that the potential of the land was not fully realised. If one farmer increases his number of cattle then he will benefit from the value of the extra cows and their milk. The costs of exceeding maximum sustainable yield and the resulting resource depletion would be shared by all the farmers. Hardin's 'tragedy' is that all the farmers would feel bound to add extra animals: if the individual did not, others would without he or she sharing the benefits. The result is the common rapidly degrades.

Although simplistic, the Tragedy of the Commons does explain the tendency to over-exploit shared resources and the need for agreements over common management.

THE COMMON FISHERIES POLICY

The Common Fisheries Policy (CFP) or **'Blue Europe'** was designed to create free trade in fish inside the Union and also to conserve fish stocks. The conservation element of the CFP combines several different strands:

1. **Technical measures** including gear regulations, minimum landing sizes, catch restrictions, closure areas.
2. **Structural reform measures** intended to reduce overall fishing capacity.
3. The setting up of **Total Allowable Catches** (TACs) for each stock, covering species in a given area.

In theory, TACs are the most comprehensive form of management. Levels are assessed annually by fisheries' biologists and then divided into national quotas according to specific areas. When a TAC or resulting quota has been exhausted, the fishery must be closed.

Of all the CFP's measures the TACs have caused the most trouble. The allocation of resources among member states in the form of national catch quotas has led to arguments. Other problems include fraud by fishermen, poor policing, overestimation of fish numbers and the fact that some TACs have been set above levels suggested by scientists. In 1993, for example, the Commission set higher ceilings than those recommended for catches of hake in Spanish and Portuguese waters, sole in the North Sea, saithe off the West of Scotland and whiting in the Irish Sea.

The failure of 'Blue Europe'

It seems that after more than ten years of fisheries policy the EU's fishermen still have problems in catching fish economically and efficiently. The outlook for fish stocks is bleak and continuing bad management has put pressure on wildlife.

The reasons for this apparent failure are complex but are associated with a lack of competitiveness and productivity in the industry. The sections of the EU fleet which suffer from overcapacity, high indebtedness, distance from fishing waters and high operating costs have been the worse hit. However, the industry as a whole is still in a crisis based on a combination of both external and internal factors:

- **External factors**: competition from non-EU countries, agricultural alternatives to fish such as low-to-mid-price meat products, and aquaculture (fish farming).
- **Internal factors**: remaining overcapacity, overfishing by fishermen with little faith in governments or the scientists' quotas, unequal policing and monitoring of quotas, fraud and misrepresentation of catch size.

Common cheats include false reporting of catches, double-hulled boats to hide extra fish and catching juvenile fish. Fraudulent catch in some EU waters is at least 10-20% over the limit with as much as 40% of the catch being discarded.

Despite the efforts of the CFP to restrain levels of fishing, many of the most important commercial species are suffering from excessive rates of exploitation:

1 Spawning stocks are seriously depressed.
2 Yields from the fishing grounds fluctuate markedly about a declining trend.
3 Landings from the North Sea are falling.

Figure 2.20 shows just how badly hit cod has been by overfishing. Experts recommend an end to North Sea cod fishing. Stocks of plaice, sole and northern hake in the waters off south west England are also close to, or beyond, the biological limits which determines sustainability.

The future

The CFP has been unpopular with both environmentalists and fishermen. In the case of fishermen the results have been quite dramatic. In early 1993 and 1994, French fishermen staged violent protests against cheap fish imports, extracting emergency measures. In summer 1994, Spanish trawlermen clashed in the Bay of Biscay with British, French and Irish vessels, which they accused of using oversized nets. The latest cause of dissatisfaction with the CFP is a deal agreed in December 1994 over the access of the large Spanish fleet to two protected and vulnerable waters off the UK and Ireland: the 'Celtic Sea', between south east Ireland and south west England and the 'Irish Box', the coastal waters around Ireland.

This has led to inevitable criticism by fishermen in the UK but also in Spain because access was blocked to the valuable waters of the Bristol Channel and the Irish Sea.

In 1994, the European Commission responded to this militancy and to the more general criticisms by publishing 'The New Common Fisheries Policy'. It suggests a more efficient monitoring of catches based on these proposals:

1 Computer databases to follow the movement of fish from producer to consumer by logging catches, landings, transport and sales.
2 Satellites used to monitor and track boats.
3 Penalties strengthened and standardised between and within the member states.

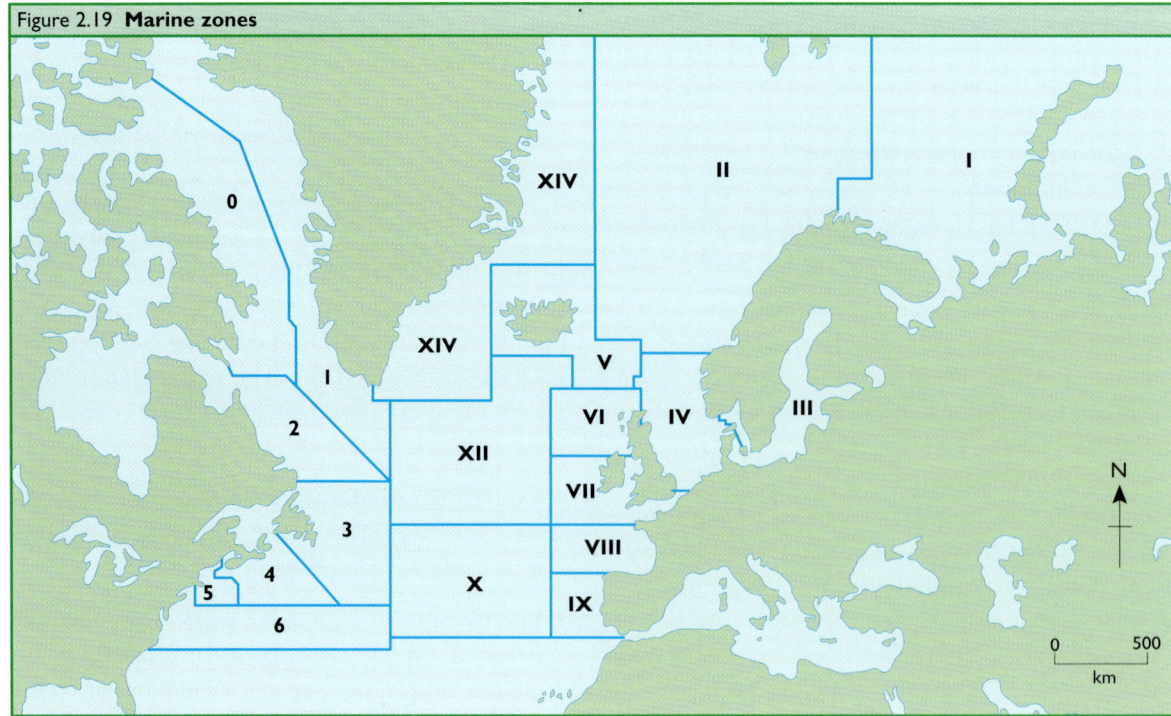

Figure 2.19 **Marine zones**

Too many fishermen, too few fish

Many argue that these measures still fail to address the real problems of the European fishing industry: too many fishermen are chasing too few fish and too many immature fish are being caught. For the fisheries to be protected and for the industry to be competitive on a world scale, the number of boats and the number of men employed in fishing must be reduced. At the same time, the efficiencies which come from improved technology must be embraced. **Figure 2.21** suggests some possible strategies for the future but there are clearly no simple solutions to the problems associated with such a politically, economically and environmentally sensitive industry.

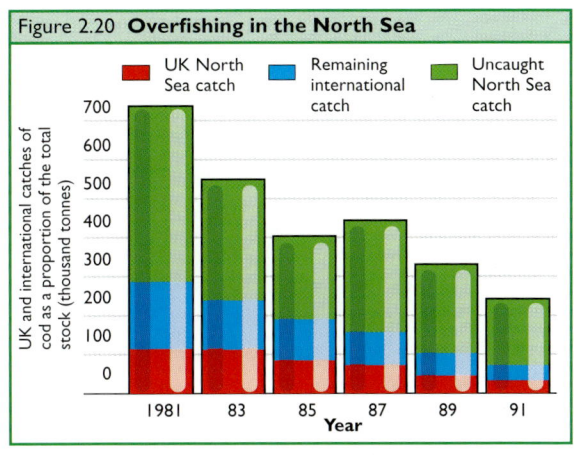

Figure 2.20 **Overfishing in the North Sea**

Figure 2.21 **Strategies for the European fishing industry**

ACTION	TYPE OF MEASURE	OBJECTIVES
Conservation of resources		
Technical measures	Small meshed nets, minimum landing sizes, boxes	Protect juveniles and encourage breeding, discourage marketing of illegal catches
Restrict catches	TACs and quotas	To match supply to demand, plan quota uptake throughout the season, protect sensitive stocks
Limit number of vessels	Fishing permits (which could be traded inter- or intra-nationally)	System applicable to EU vessels and other countries' vessels fishing in EU waters
Surveillance	To check landings by EU and third-country vessels (log books, computer/satellite surveillance)	To apply penalties to overfishing and illegal landings
Structural	Structural aid to the fleet	Finance investment in fleet modernisation (although commissioning of new vessels must be closely controlled) whilst providing reimbursement for scrapping, transfer and conversion
Reduction in employment leading to an increase in productivity	Inclusion of zones dependent on fishing in Objectives 1, 2 and 5(b) of Structural Funds	To facilitate restructuring of the industry, to finance alternative local development initiatives to encourage voluntary/early retirement schemes
Markets		
Tariff policy	Minimum import prices, restrictions on imports	To ensure EU preference (although still bound under WTO)
Other measures		
Restrict number of vessels	Fishing licenses	Large license fees would discourage small, inefficient boats
Increase the accountability of fishermen	Rights to fisheries	Where fish stay put (e.g. shellfish) sections of the seabed can be auctioned off
		Where a whole fishery is controlled, quotas could be traded which would allow some to cash in and leave the sea

Section C Exercises and recommended reading

EXERCISES

1 Define the following terms: *extensification, specialisation, fragmentation, price support, marginal farms.* [5]

2 Study **Figure 2.22** which shows wheat yields in Europe for 1971 and 1992.
   (a) Describe the pattern of wheat yields for 1971. [4]
   (b) How does the pattern of wheat yield in 1992 differ from that of 1971? [5]
   (c) Explain the variations in wheat yields in Europe with respect to:
       (i) the growing requirements of wheat
       (ii) the physical geography of Europe
       (iii) access to the main urban markets [12]
   (d) How useful are the models of agricultural geography in explaining
       (i) the distribution of wheat production
       (ii) changes in the distribution of wheat production from 1971–1992 [7]
   (e) Describe and explain the likely environmental consequences of the increases in wheat production since 1971. [5]
   (f) Evaluate the two ways of illustrating data as shown in **Figure 2.22**. What do you think are the strengths and weaknesses of each method and which do you think is most appropriate for showing variations in wheat production. Justify your answer. [12]

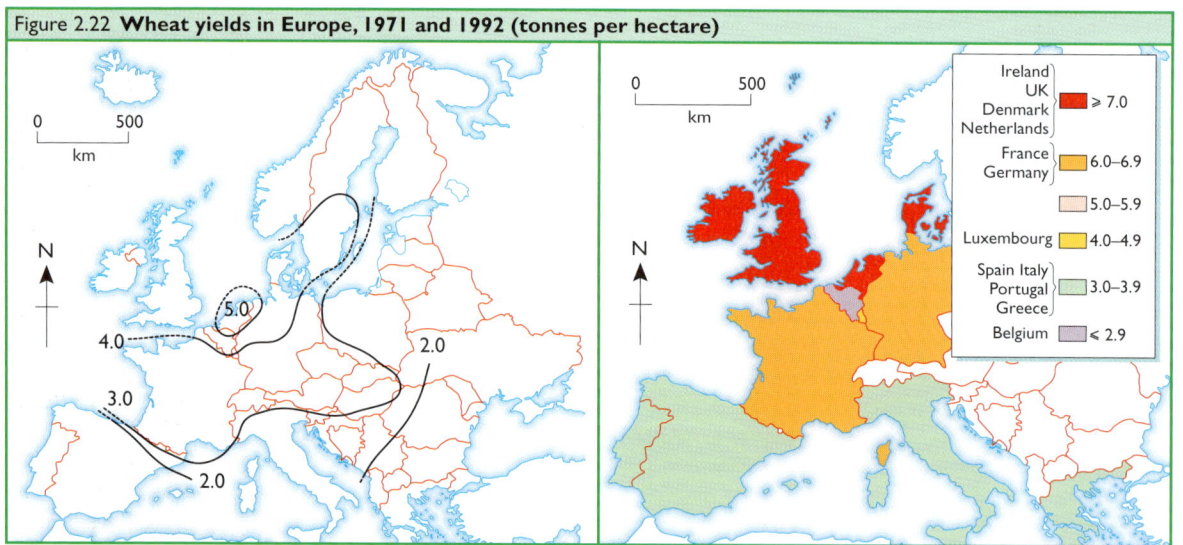

Figure 2.22 **Wheat yields in Europe, 1971 and 1992 (tonnes per hectare)**

RECOMMENDED READING

Useful sources include MAFF's *Agriculture in the UK* (1995), *Regional Trends 30* (1995), the EU's *Our Farming Future* (1993). Good textbooks on agriculture include Ilbery's *Agricultural Change in Great Britain* (Oxford, 1990) and Grigg's *An Introduction to Agricultural Geography* (Hutchinson, 1984). Most textbooks on Europe (see p. 18) contain a section on agriculture though not all have fishing. Articles on the environmental impact of agriculture include: 'The Nitrate Issue' (1989) in *Geography Review*, vol. 2, no. 5, pp. 28-31, 'Water erosion of arable land on the South Downs' (1990) in *Geography Review*, vol. 4, no. 1, pp. 19-23 and 'Damage to property by run-off from agricultural land, South Downs, southern England 1976-93' (1995) in *The Geographical Journal*, vol. 161, no. 2, pp. 177-91. Bowler's article in *Geography* (1986), pp. 14-24 is an excellent starting point on regional changes in European agriculture.

CHAPTER 3
FACTORS AFFECTING ECONOMIC AND REGIONAL DEVELOPMENT

A **INDUSTRIAL CHANGE**
  Structure and location..**42**
  Deindustrialisation and reindustrialisation..**44**
  Rationalisation and restructuring...**45**
  Small and medium sized enterprises...**46**
  National and international regional aid...**48**

B **CLASSICAL LOCATION THEORY**
  Traditional industrial location models...**50**

C **NEW LOCATION THEORY**
  Spatial division of labour...**52**
  Globalisation and glocalisation..**55**
  Just-in-time and flexible production..**56**
  Industry and the environment...**57**

D **EXERCISES AND RECOMMENDED READING**..**58**

Manufacturing industry is the transformation of raw materials into new products. Increasingly, in developed economies it involves complex methods and specialisation of labour although it employs a declining proportion of the workforce.

Section A shows that manufacturing industry has a very dynamic and technical nature characterised by increasing efficiency, output, complexity and technology, as well as by changes in demand, raw materials, and in the location of production sites. For Geographers key questions regarding industry include 'Why are certain regions attractive for specific industry?' and 'What acquired advantages have they accumulated as a result of their development?'

The location of manufacturing industry depends on a wide variety of influences. These include a number of factors such raw materials, energy resources, transport costs, markets, labour supply and availability including skilled management, capital, government incentives and human behavioural factors.

Sections B and C investigate the changing influence of these factors in a European context and analyses models of industrial location. It is difficult to assess accurately the importance of manufacturing industries. A number of indices can be used, none of which is entirely satisfactory. The number of factories, size of workforce, percentage of total workforce, level of mechanisation, capital investments, machinery value, physical output, type and quality of manufacturing can all be used. If there are limitations in assessing the importance of manufacturing there is as much, if not more, difficulty in delimiting manufacturing regions. The Ruhr, the Midlands and the Nord in France are identified as manufacturing areas yet much of their manufacturing employment and industries have been declining for many decades. It is a complex pattern.

Section A Industrial change

STRUCTURE AND LOCATION

A region's **structure** refers to the resources it has at its disposal, including raw materials and energy reserves. **Location** refers to the relative ease of access and communications that an area possesses. The importance of raw materials, energy resources and communications is examined in greater detail in **Chapter 4**; here, they are considered in a more general sense. Structure and location are fundamental in explaining regional prosperity and changes in prosperity over time. For example, an economy based on high-technology industry and services, such as the South East in the UK or the Rhone-Alpes in south west France, has a very favourable economic structure and a good location relative to Europe. These areas are more likely to grow and become core regions. By contrast, areas which have a relatively high proportion of the workforce employed in out-dated industries, such as shipbuilding in the North East of England or iron and steel in southern Italy, have a poor economic structure as well as having a poor location, i.e. they are peripheral to the main markets.

Matrix of regional prosperity

These two factors can be combined to provide a simple grid which assesses the potential for regional prosperity (**Figure 3.1.**) This matrix can be examined at a variety of scales. **Figure 3.2** shows that the **core areas** of Europe are relatively central to the main centres of population in Europe and have good communications networks. They are also close to centres of decision making and government, vital for the day to day contacts needed in service industries.

However, structure and location are relative and can **change over time**. For example, within a core region there may be peripheral areas and vice versa. Although the South East of England is widely recognised as forming part of Europe's core, the Thames Gateway is a depressed region associated with an out-dated economic structure (port industries) and relatively poor communications. By contrast, in the North West, Birchwood Science Park is an excellent example of a growth pole with a structure based on high technology, chemicals and engineering and a location close to the intersection of the M6 and M62 motorways. Birchwood is within about two hours drive of about one-third of the UK population, a market of nearly twenty million people, and its proximity to Manchester airport gives it greater access to the European market of some 370 million people.

Moreover, economic structure changes as the **utility of natural resources** change. One of the best examples of a changing structure is the changes in energy resources. Cheap, plentiful and reliable supplies of energy were responsible for industrial and regional development in Europe and even in the 1990s a map of oil and gas resources indicates development potential just as the distribution of coalfields is largely a map of deindustrialisation. As resources change, because of availability or competition, so too does industrial and regional potential (**pp. 60–3**).

Figure 3.1	**Potential for regional prosperity**		
	LOCATION		
STRUCTURE	**CENTRAL**	**MARGINAL**	**REMOTE**
VERY GOOD	CORE London Randstad	SECONDARY CORE Manchester Rhone-Alpes	GROWTH POLE Eastern Scotland (oil) Taranto
MEDIUM	SECONDARY CORE Northern Italy South East England	TRANSITIONAL East Midlands Denmark	INNER PERIPHERY Northern Ireland Ireland
POOR, OUT-DATED	GROWTH POLE POTENTIAL Thames Gateway Ruhr	INNER PERIPHERY North West England Alsace-Lorraine	OUTER PERIPHERY Highlands and Islands (Scotland) Mezzogiorno

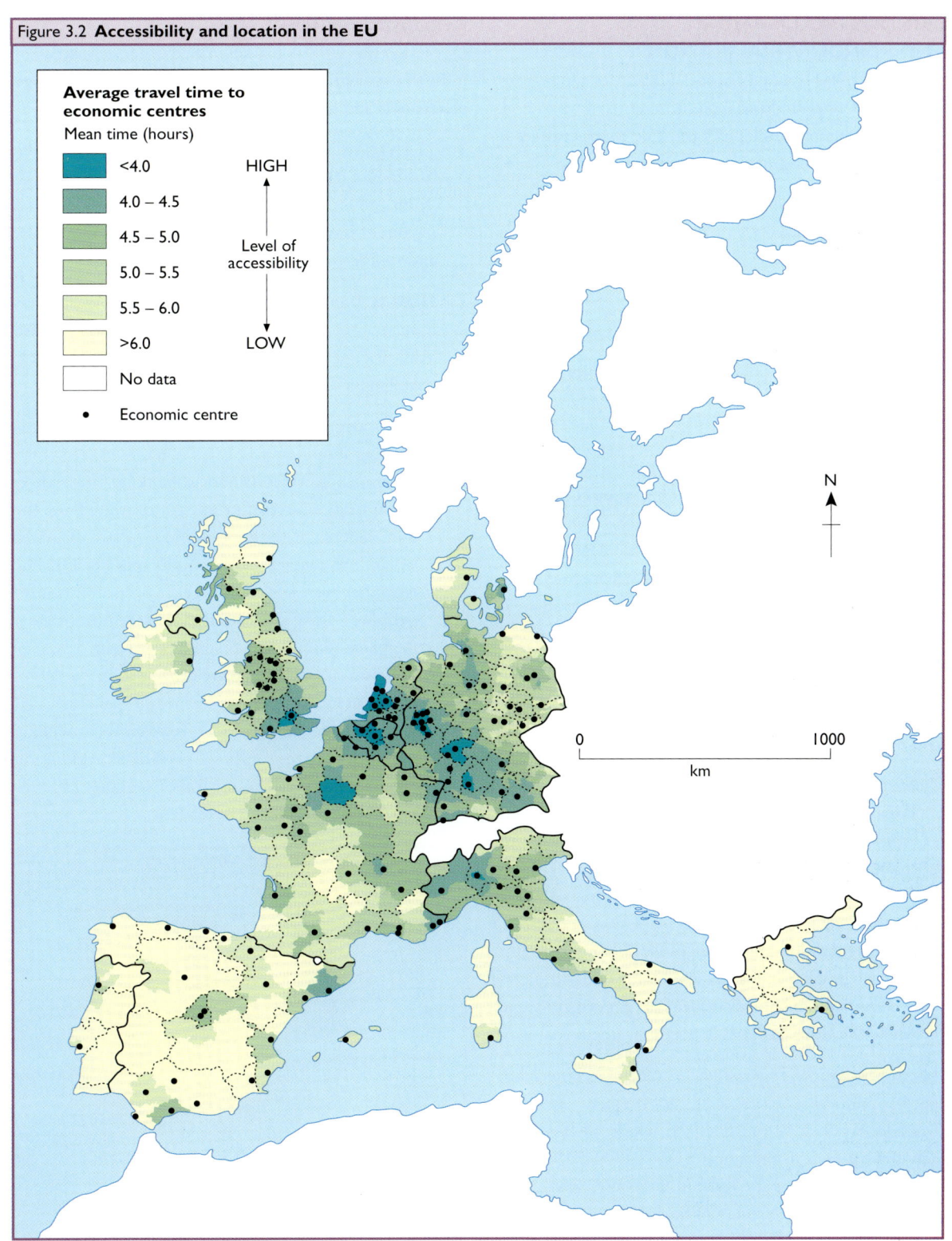

Figure 3.2 Accessibility and location in the EU

DEINDUSTRIALISATION AND REINDUSTRIALISATION

Since World War II many European countries and regions have entered a new industrial phase in terms of what, how and where manufacturing produces. The characteristics of this so-called post-industrial society are:

- A shift away from agriculture and manufacturing industries and towards service industries, especially in terms of employment.
- A trend towards large corporations characterised by diversification, mergers and joint ventures.
- An international division of labour.
- The decline of the older smokestack industries which grew up in the nineteenth century.

Deindustrialisation

The term deindustrialisation refers to a long term, absolute decline of employment in the manufacturing sectors. It must be stressed that it refers to a loss of jobs rather than a decline in productivity. A key element of **post-industrial** Europe is contained in the concept of deindustrialisation. The decline of certain industries or areas is due to a number of factors: exhaustion of resources; increasing costs of raw material; automation and new technology; introduction of a rival product; fall in demand; overseas competition; rationalisation; a rise in costs; the removal of a subsidy and a lack of capital.

There are two types of deindustrialisation:

1. **Positive deindustrialisation** occurs when industries reduce their workforce to increase productivity; this model sees an improvement of competitiveness through mechanisation and rationalisation and where displaced labour is absorbed in the service sector.
2. **Negative deindustrialisation** occurs when the decline affects particular industries and regions due to inefficiencies in production or inadequate infrastructure and where the displaced labour results in unemployment.

Deindustrialisation is concentrated in the EU's nineteenth century industrial cities such as Bremen, Utrecht, Tyneside, Strathclyde and Sheffield which all lost more than 20% of their manufacturing base between 1973 and 1983. Regions such as the Ruhr in Germany, Sheffield in the UK and the Bilbao region of Spain have all attempted to reindustrialise through a series of regional policies.

Reindustrialisation

David Keeble (1989) has identified three key areas of industrial growth: **small firm growth**, **high-technology electronics** and **services** (tertiarisation):

1. The highest rates of **small firm** growth are found in the less industrialised, rural, peripheral areas such as the 'Midi' of southern France, the Adriatic and central regions of Italy, East Anglia and rural Wales in the UK, and southern Germany bordering the Alps.

2. **High-technology firms** have grown up in the less industrialised 'sunbelt' regions such as the Cote d'Azur, Toulouse, Grenoble, Bavaria, Baden-Wurtenburg, East Anglia, Devon and Somerset.

3. **Tertiarisation** has two key aspects, high level producer services and tourism:
   - **Producer services** (finance, banking, insurance, marketing, advertising) are concentrated on the state capitals of the EU—London, Paris, Brussels, Milan, Amsterdam and Frankfurt (**pp. 88-93**). The result has been some convergence of specific centre-periphery disparities with regard to the nine original members (not Greece, Spain or Portugal).
   - **Tourism** has seen a massive growth since the 1970s due to an increase in personal income, paid leave and package holidays (**pp. 96-7**). Tourism is particularly important in the unspoilt peripheral regions: in Spain tourist numbers rose from 7 million to 33 million between 1960 and 1980.

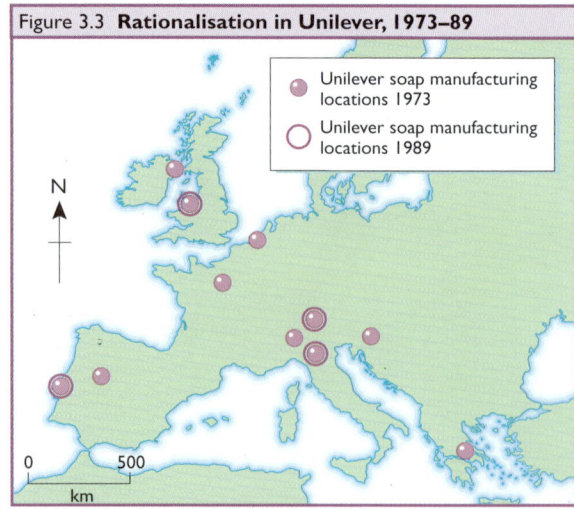

Figure 3.3 **Rationalisation in Unilever, 1973–89**

RATIONALISATION AND RESTRUCTURING

The concepts of **globalisation** and **spatial division of labour** suggest a **deconcentration** and **decentralisation** of manufacturing industry at the national, European and world scales. These changes are also associated with restructuring and closure. An example from the chemicals sector can be seen in **Figure 3.3**. This shows the **rationalisation of Unilever** toilet soap manufacture in Europe caused by the run-up to the single European Market of 1992. Its manufacturing base has been rationalised from eleven factories in the EU down to four between 1973 and 1989. Restructuring within Europe toward a single European Market has also influenced the international activities of Japanese firms in two ways:

1. Companies not already located in Europe have moved into the EU to obtain a direct manufacturing presence. This includes automotive companies such as Toyota, Nissan and Honda who have established key manufacturing plants in the UK in order to gain direct access to European markets.

2. Japanese firms with an existing presence within Europe have sought to gain benefits from rationalisation.

Thus, **Toshiba** with some 87 companies in 11 European countries has undergone restructuring and closure of some of these production facilities in order to achieve greater manufacturing efficiencies and more effective Europe-wide coordination. The decision by **Hoover** in 1994 to close its plant at Dijon and concentrate production at its other plant in Glasgow is another highly controversial example.

Recession

Another factor affecting the rationalisation of firms from the 1980s onwards has been the recession. This has been felt by many international companies and industrial sectors:

- **The steel industry:** the European steel industry has been left with an overcapacity of over 30 million tonnes and this has meant plant closures throughout Europe including Ravenscraig in Scotland and Consett in North East England.

- **Electronics and computers:** the largest companies have been hardest hit. IBM rationalised its world operations cutting employee numbers from 382,000 in 1989 to 344,000 by the mid-1990s. Alongside this the company has re-organised into 13 semi-autonomous business units.

- **Automotive industry:** the recession has led to a fall in sales worldwide leading to increasing conflict between Japanese, US and EU producers. The Japanese have expanded in Europe and GM are rationalising in favour of Europe by closing plants in the USA.

Rationalisation can be considered a natural part of the process of globalisation. The result is that many of Europe's regions are in competition not only to attract investment but also to attract further reinvestment.

Figure 3.4 shows a possible model for the global expansion of a company. It starts from a point where a company begins to export from its home base through location overseas and finally a decision to rationalise its global location in areas of comparative, competitive or strategic advantage.

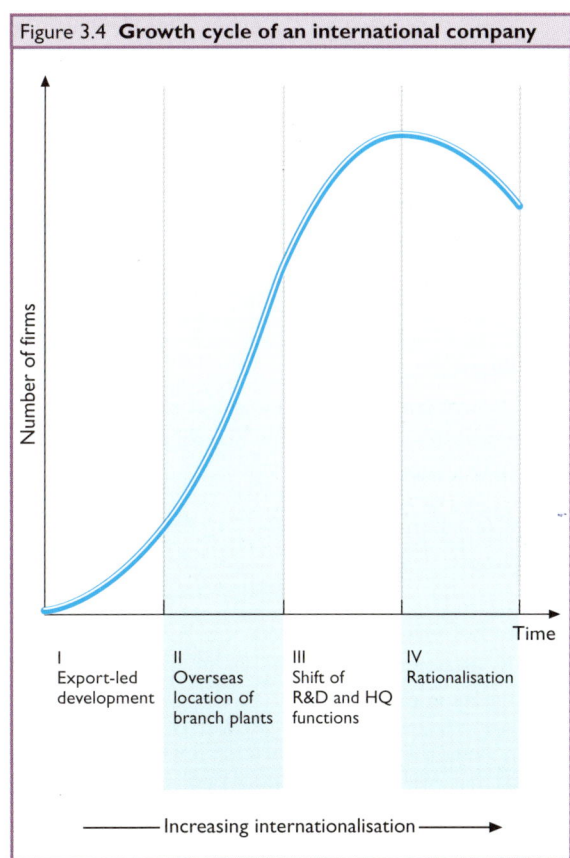

Figure 3.4 **Growth cycle of an international company**

1. **Export-led development** has the advantages of concentrating activities in the home country where labour and sourcing are established. However, the exports may be subject to tariffs or trade restrictions.

2. **Overseas location** of branch plants has the advantage of overcoming trade restrictions and accessing cheap labour, financial assistance and new markets. Many companies remain in this stage of internationalisation.

3. **Shift of R&D and HQ functions** shows a more fundamental commitment to globalisation. Products are not only sourced and assembled in the new markets but the new area becomes semi-autonomous. Managerial and research decisions shift away from the home country which enables a product to be shaped to the new market.

4. **Rationalisation** sees the internationalised company concentrating activities in the **best locations**. Some profitable plants may be closed down or downgraded because their activities are represented elsewhere in lower cost or more strategic locations.

SMALL AND MEDIUM SIZED ENTERPRISES

Small and medium sized enterprises (SMEs) include three main categories:

1. Self-employed persons.
2. Micro-sized firms employing less than 10 people.
3. SMEs proper, employing between 10 and 499 people.

Significance

SMEs are extremely important to the economy of Europe both industrially and regionally. The following statistics, from 1992, make this clear:

- 90% of all enterprises in the EU have less than 10 employees.
- only 13,000 of the 14.6 million enterprises in the EU have more than 500 employees.
- SMEs account for 52.4 million of the 88.5 million employees in the EU.

Within these general statistics there are important industrial and regional differences within the EU:

- **Industrial:** construction as well as trade and service industries are dominated by small firms; larger firms are much more important in mining and manufacturing industries.

- **Regional:** a high SME presence is a feature of peripheral regions. However, in reality, the situation is more complex. SMEs are not a homogeneous class but rather differ strongly in size, technological competence, market orientation, organisation and linkages. These differences are reflected in the location of different types of SMEs within the EU.

Classification

SMEs have been divided into five categories based on industrial, locational and technological criteria:

1. **Market localists** operate exclusively in local markets serving local customs, tastes and traditions. They operate in the fields of construction, food processing, clothing and furniture. One example is the brewing industry in Bavaria where the small-scale producers serve local tastes. The entrepreneurs often lack formal training and are rooted in family or neighbourhood ties. They are more important to the economies of peripheral countries.

2. **Craft-based enterprises** produce diversified and customised goods of high quality which may be aimed at market niches at the regional, national or even international level. In Germany and Belgium the strong representation of Chambers of Craft and Chambers of Industry and Commerce, linked with other institutions such as local banks and vocational schools, has led to the continued strength of craft industries.

3. **SMEs within regional networks** (industrial districts) have elements of market localists and craft-based SMEs but operate in production clusters with close ties between each other and also with regional governments and unions. This model allows for specialisation and market strength through cooperation. These agglomerative clusters of SMEs have been particularly successful in Baden Würtenburg in Germany, Jutland in Denmark and in the so-called 'Third Italy' concentrated in the North Central and North East of Italy (centred in the Emilia Romagna region based on the regional capital, Bologna).

4   **High-tech SMEs** (technological districts) are also clusters of SMEs but with a specific high-tech orientation. Under a diversity of headings such as Science Parks, Technology Parks, Innovation Centres and Technopoles, 'technological districts' have emerged in different parts of Europe to offer opportunities to knowledge-based SMEs. The French technopoles in Grenoble and Sophia-Antipolis and the M4 corridor between London and Bristol are good examples.

5   The **division of labour** between **SMEs and large firms** has links to the success of the technological districts. Many SMEs are used as sub-contractors for standardised parts, components and research in the computer and car industries. This can be seen in the growing relationship between British SMEs and the Japanese transplants in car and consumer electronics. In regions like the industrial district of Baden-Wurtenburg, in Southern Germany, there are even stronger arrangements between large and small firms in the automobile, electrical engineering and textile industries.

Kingdoms and republics

Traditional wisdom holds that SMEs are economically successful because they do things that large firms are unable to do or cannot do efficiently, for example:

- Serving local markets and market niches.
- Producing highly specialised goods.
- Exploiting small local labour forces.

R. Rothwell (1986) has suggested a dynamic relationship between small firms and large firms (**Figure 3.5**). However, large firms are becoming increasingly flexible developing towards systems of firms-in-the-firm (in which a large firm is broken up into competing management or production units) with similar flexibility and capabilities to SMEs. If SMEs are to compete with large firms rather than be engulfed or left stranded by them they must match the large resources of the major companies. Two strategies have been suggested:

1   '**Kingdoms**': SMEs work closely together with one large sized firm which provides market access, strategic planning, security and guidance on research and development, e.g. SMEs in the new division of labour (class 5, above).

2   '**Republics**': SMEs collaborate on an equal footing and bring their particular strengths to products and the production process allowing for specialisation, a sharing of costs and marketing strength, e.g. high-tech SMEs (class 4, above) and SMEs within regional networks (class 3, above).

Figure 3.5 Advantages *(italic type)* and disadvantages *(bold type)* of small firms and large firms		
	SMALL FIRMS	**LARGE FIRMS**
MARKETING	*Ability to react quickly to keep abreast of fast changing market requirements.* **Market start-up abroad can be prohibitively costly.**	*Comprehensive distribution and servicing facilities. High degree of market power with exisiting products.*
MANAGEMENT	*Lack of bureaucracy. Dynamic, entrepreneurial managers react quickly to take advantage of new opportunities and are willing to accept risk.*	*Professional managers able to control complex organisations and establish corporate strategies* **Can suffer an excess of bureaucracy. Often controlled by accountants rather than by entrepreneurs. Managers can become mere 'administrators' who lack dynamism with respect to new long-term opportunities.**
INTERNAL COMMUNICATIONS	*Efficient and informal internal communication networks. Fast response to internal problem-solving. Provides ability to reorganise rapidly to adapt to change in the external environment.*	**Internal communication often cumbersome. This can lead to slow reaction to external threats and opportunities.**
QUALIFIED TECHNICAL MANPOWER	**Often lack qualified technical specialists and unable to support a formal R&D effect on an appropriate scale.**	*Ability to attract highly skilled technical specialists. Can support the establishment of a large R&D laboratory.*

NATIONAL AND INTERNATIONAL REGIONAL AID

Regional policy in the UK

Governments become involved with regional problem areas for a variety of social, economic, environmental and political reasons (**Inset 3.1**), notably above average rates of unemployment. Successive governments have attempted to reduce regional disparities by attracting new industrial development and investments with a variety of inducements such as low rents, tax relief, advanced factories, relocation grants and labour subsidies. Nevertheless, despite over sixty years of regional policy, inequalities still exist and the pattern has remained quite persistent until the late 1980s.

During the 1980s the North-South divide in Britain intensified, with the North increasingly characterised by deindustrialisation compared with service growth and private initiative in the South. However, by the end of the 1980s and through the 1990s Britain experienced its most severe depression since the war. Much of the effects were felt in the more prosperous South, especially in the service sector and the defence-linked industries. Between 1990 and 1992 the South East experienced the greatest percentage rise in unemployment, declining output, above average decline in house prices, and high rates of bankruptcies. The government's revised regional assistance policy was produced in 1993 and this took into account declining conditions in areas not traditionally associated with regional problems such as Thanet, Dover and Deal, Folkestone and the Isle of Wight. Regions were selected on the basis of a number of factors: current, long-term and/or structural unemployment, demographic structure, labour demand, activity rates, peripherality and imminent changes in local job market. On this basis, two tiers were identified: **Development Areas** (DAs) and **Intermediate Areas** (IAs), covering 16% and 18% of the population. DAs receive more help than IAs. For example, DAs can receive up to 30% of the costs of a new development project, compared with 20% in IAs. **Regional Enterprise Grants** (REGs) are available to all small firms in DAs but only some in IAs, and there is greater government funding in DAs for each new job created.

INSET 3.1 REGIONAL ASSISTANCE

Why do governments get involved?
1 Economic reasons such as reducing unemployment, utilising resources and raising productivity.
2 Social reasons such as reducing income inequalities, stemming migration and raising standards of living.
3 Political manipulation especially in the run-up to an election.
4 Strategic reasons in times of military aggression or threats to national security.
5 Environmental concern over contamination, dereliction and blight.

What are the aims of regional policy?
1 Encouragement of development and readjustment.
2 Control of growth.
3 Reduction of regional inequalities.

How does regional aid come about?
1 Public sector investment in infrastructure and human resources.
2 Locational intervention through inducements and restrictions.
3 Individual agreements such as compulsory purchase orders, compensation and so on.

Regional policy in the EU

The EU has developed a number of programmes to combat regional inequalities by improving regional infrastructure and increasing the standards of living in deprived areas. Of these, three have a regional expression: Objectives 1, 2 and 5b—less developed regions, declining industrial regions and rural problem regions respectively. The others, Objectives 3 (long-term unemployment), 4 (young people) and 5a (farm modernisation) are not defined on a regional basis.

Objective 1 regions are those that are lagging behind the progressive regions. These have a GNP of 75% of the EU average and include the Scottish Highlands, Northern Ireland, Ireland and the Mezzogiorno. These are at the centre of the

INDUSTRIAL CHANGE **49**

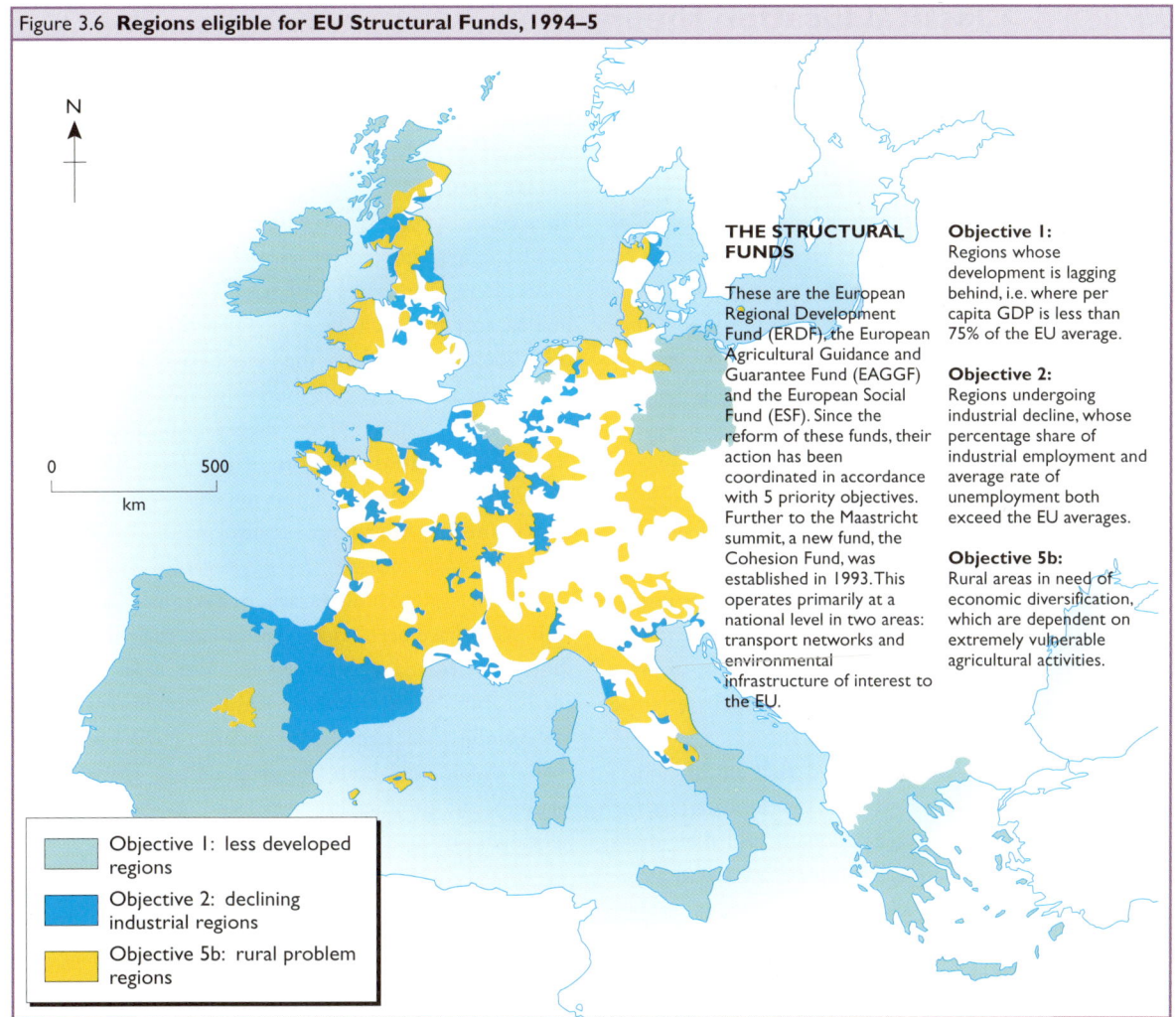

Figure 3.6 **Regions eligible for EU Structural Funds, 1994–5**

THE STRUCTURAL FUNDS

These are the European Regional Development Fund (ERDF), the European Agricultural Guidance and Guarantee Fund (EAGGF) and the European Social Fund (ESF). Since the reform of these funds, their action has been coordinated in accordance with 5 priority objectives. Further to the Maastricht summit, a new fund, the Cohesion Fund, was established in 1993. This operates primarily at a national level in two areas: transport networks and environmental infrastructure of interest to the EU.

Objective 1: Regions whose development is lagging behind, i.e. where per capita GDP is less than 75% of the EU average.

Objective 2: Regions undergoing industrial decline, whose percentage share of industrial employment and average rate of unemployment both exceed the EU averages.

Objective 5b: Rural areas in need of economic diversification, which are dependent on extremely vulnerable agricultural activities.

- Objective 1: less developed regions
- Objective 2: declining industrial regions
- Objective 5b: rural problem regions

EU's plan to reduce regional inequalities, and up to 74% of the EU's annual structural budget (ECU 22.7 billion in 1996) is directed towards Objective 1 areas. Development strategies centre on infrastructural improvement—transport, communications, services and facilities.

Objective 2 areas are those associated with the decline of traditional industries, e.g. coal, iron and steel, textiles and shipbuilding. These are characterised by above average rates of unemployment and an outdated manufacturing sector, for example much of South Wales, the Thames Gateway and the Ruhr coalfield. Emphasis is placed on job creation, renovation and reconstruction. New small and medium sized enterprises (SMEs) are being targeted as potential new investors. However, only six per cent of structural funds are available to Objective 2 areas.

Rural problem regions make up **Objective 5b**. Attention is focused on non-agricultural improvements, e.g. tourism, transport, service industries and the provision of amenities. Halting rural depopulation is a central aim. Five per cent of funds are available for Objective 5b. These and other schemes orientated to agricultural areas have been examined on **pages 34–5**.

Other policies have a regional impact. The EU's ECU 300 million initiative to support peace in Northern Ireland and the EU's aid for coal and steel areas, 1994-7, are good examples. Under the latter, the East Midlands will receive over ECU 40 million and Wales over ECU 20 million from Structural Funds for regeneration.

Section B Classical location theory

TRADITIONAL INDUSTRIAL LOCATION MODELS

Early attempts at explaining the location of manufacturing tended to stress the historical evolution of the industry, especially with regard to the distribution of physical factors such as the availability of energy resources, raw materials, water and cheap flat land. As geography developed during the twentieth century a greater preoccupation with models became evident and industry was no exception. At least four groups of models can be identified: urban-industrial location, least cost location, spatial margins and behavioural approaches.

Models of urban land use have located manufacturing industry in inner city areas (**Burgess's concentric model**), along major routeways (**Hoyt's sector theory**), and in industrial suburbs (**Harris and Ullman's multiple nuclei model**). These reflect the variety of manufacturing industries and their differing locational requirements. In these models the location of industry is described but little explanation is given why it is there.

In 1909 Alfred **Weber** predicted that industrialists would locate their factory at the least cost location, i.e. the cheapest site, since they acted rationally and their main aim was profit maximisation. His model was based mostly on transport costs but also took into account labour costs and agglomeration economies (the savings that could be made through sharing). He predicted that industries that experienced significant weight loss during manufacturing, such as coal in the manufacture of steel, would locate close to the raw material whereas industries which experienced weight gain during processing, such as brewing, would locate close to their market. Geographers have long noted the shortcomings of Weber's model: industrialists are not rational, there can be only one least cost location although many profitable ones, and he over-emphasised the importance of transport costs. However, it has merit. Many MNCs seek supplies of cheap labour, not only in the developing world but in Europe too, for example the car plants at Zaragoza and Valencia in Spain. Second, iron and steel works that are found in coastal locations, such as Llanwern–Port Talbot, are in least cost locations. Moreover, the concentration of high-tech industries on sites such as the Warwickshire Science Park to share in new developments and technologies illustrate Weber's principle of agglomeration. Finally, the increasing global importance of MNCs, with their huge financial and human resources, is creating a type of rational man, intent on maximising profits and with the necessary information and backing to implement research findings.

An alternative model is that of August **Losch**. In his market area approach, he claims that firms will locate in the place that gives them the largest possible market and profit. His result is a distribution of factories similar to Christaller's central place theory. An alternative is **Rawstron's spatial margins**, which have been popularised by David Smith. These are based on areas rather than points and offer a more realistic method of analysing locational preference.

As with agricultural models **behavioural** approaches (**pp. 26-7**) have been adopted although their importance can be questioned with respect to more important structural features of the economy. Other more recent models are examined throughout this chapter. These include the spatial division of labour (the geography of 'spatial organisation'), the product life cycle, and 'globalisation' and 'glocalisation'.

Figure 3.7 **Llanwern–Port Talbot iron and steel plant**

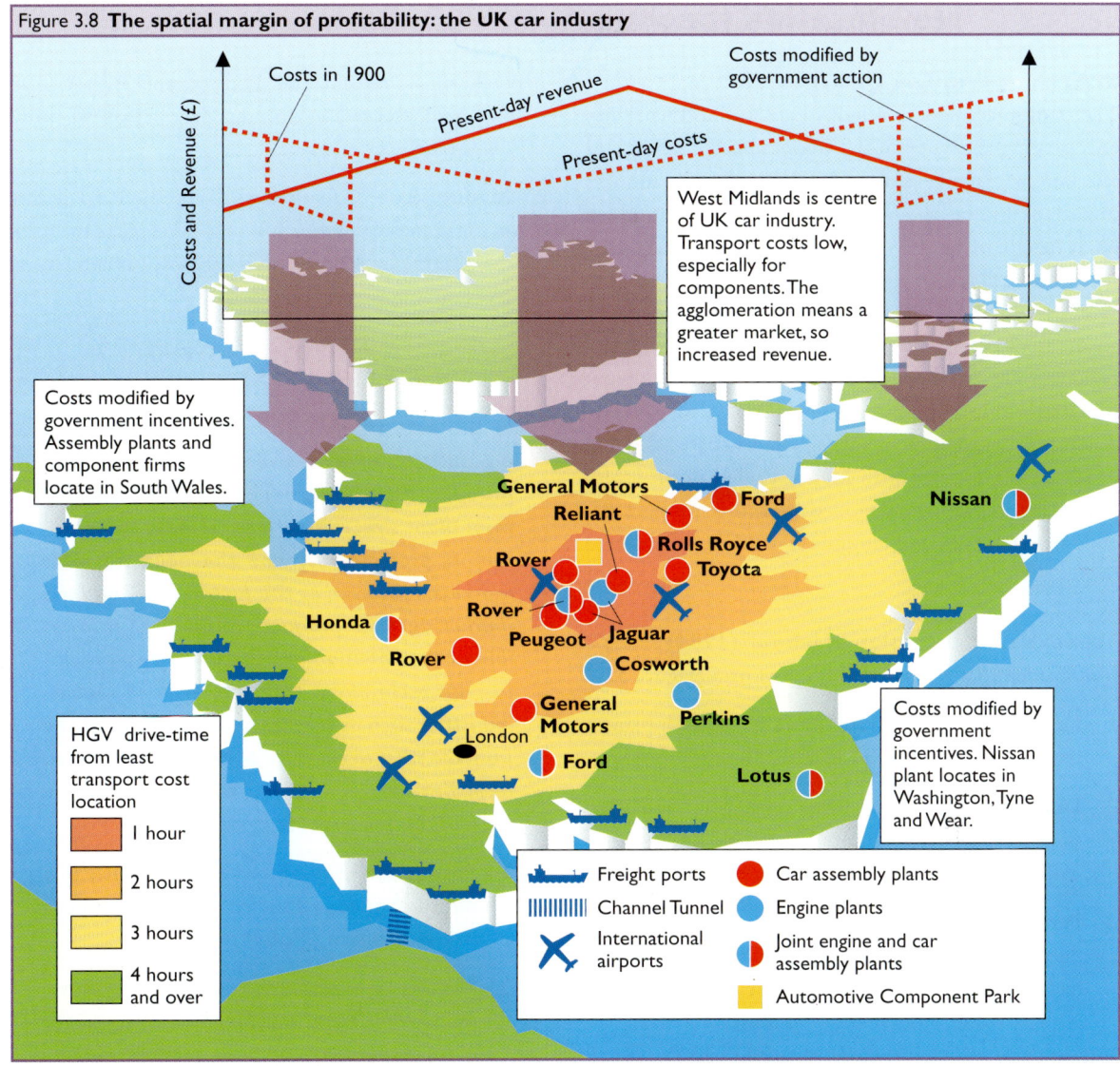

Figure 3.8 The spatial margin of profitability: the UK car industry

The spatial margins to profitability

This is a line or lines containing the area or areas within which a particular productive activity can be undertaken profitably. The spatial margin is defined by the equality of the total cost incurred and total revenue obtainable from the production of a given volume of output. Within the margin, revenue exceeds cost and a profit can be made. Beyond the margin cost exceeds revenue and production would incur a loss.

The empirical identification of the spatial margin to profitability is very difficult, although examples can be found in the literature. However, the concept is of considerable value in speculating about the likely extent of constraints on locational choice in circumstances where limited information on spatial variations in cost and revenue is available.

The theoretical significance of the spatial margin is that it focuses attention on limits to freedom of locational choice on the part of the entrepreneurs (or planners) with imperfect ability and knowledge, and away from the elusive single point where profit may be maximised.

The spatial margin was devised by E. M. Rawstron in the 1950s, and is one of the very few original concepts introduced into spatial economic analysis by a geographer.

Section C New location theory

SPATIAL DIVISION OF LABOUR

The traditional models of industrial location theory pay only passing reference to the importance of labour. As friction of distance and transport costs become less important it is the labour requirements of industry which often dictate the location or expansion of industry. At a regional and global level the concept of spatial division of labour is important. Spatial division theory suggests core areas offer skilled but expensive labour and peripheral regions a cheaper but less skilled workforce. Different activities within an industry or company are seen to have different labour requirements. The term spatial division of labour is associated with Massey (1984) but it is also important to two other models: Vernon's product life cycle model (1966) and Humphrys' model of multiplant firms (1988).

Massey's spatial division of labour (1984)

Massey identifies 3 types of multiplant firm (**Figure 3.9**):

1. **Locationally concentrated** firms where each plant is relatively self-contained and autonomous.

2. **Doming branch-plant** where ownership and decision-making is concentrated at a single headquarters with production in areas of cheaper labour.

3. **Part-processing plants** which are branch plants producing for assembly at cheaper labour location, perhaps abroad but certainly in peripheral regions.

Vernon's product life cycle model (1966)

The product life cycle model is associated with three stages of production which have clear implications both to the spatial division of labour and industrial location. In the model manufacturing capacity shifts from core areas to peripheral areas as products move into a 'mature' phase of their life cycles (**Figure 3.10**).

1. **Development phase.** Demand is small relative to the potential market and production takes place in the area of innovation which is usually a core region.

2. **Growth phase.** Demand expands rapidly as the new product becomes widely accepted. Production techniques become standardised and exports may be supplemented by foreign production to avoid tariff barriers.

3. **Mature phase.** Associated with saturation and decline in the domestic market. It is in its final phase that production moves to areas of lowest cost (usually lowest costs mean lowest labour costs).

INSET 3.2 A CRITIQUE OF THE PRODUCT LIFE CYCLE MODEL

In the past the traditional factory was built around economies of scale and the life cycle of products. Significant benefits could be gained from plant size, volume and time, by standardisation, predictability and the need for mass markets. However, as the factory ages and the market declines production moves to a lower labour cost location or a factory with more up-to-date technology. Thus, the bulk of mature industries such as textiles and steel have moved from the West to the developing world. However, new, flexible, lean production questions the logic of the product life cycle model.

1. *Production may occur simultaneously in a number of locations rather than shifting between them (global switching).*

2. *Factories are expensive to build and start up; many MNCs are unwilling to downgrade or close down plants without good reason.*

3. *Flexible production means companies are constantly embracing change within a single plant or company.*

4. *The model revolves around technological change and ignores other aspects of production — supply, demand, labour and spatial organisation.*

5. *Firms usually produce several products (and services) all with different life cycles rather than a single product with one life cycle.*

Figure 3.9 **Massey's spatial division of labour**

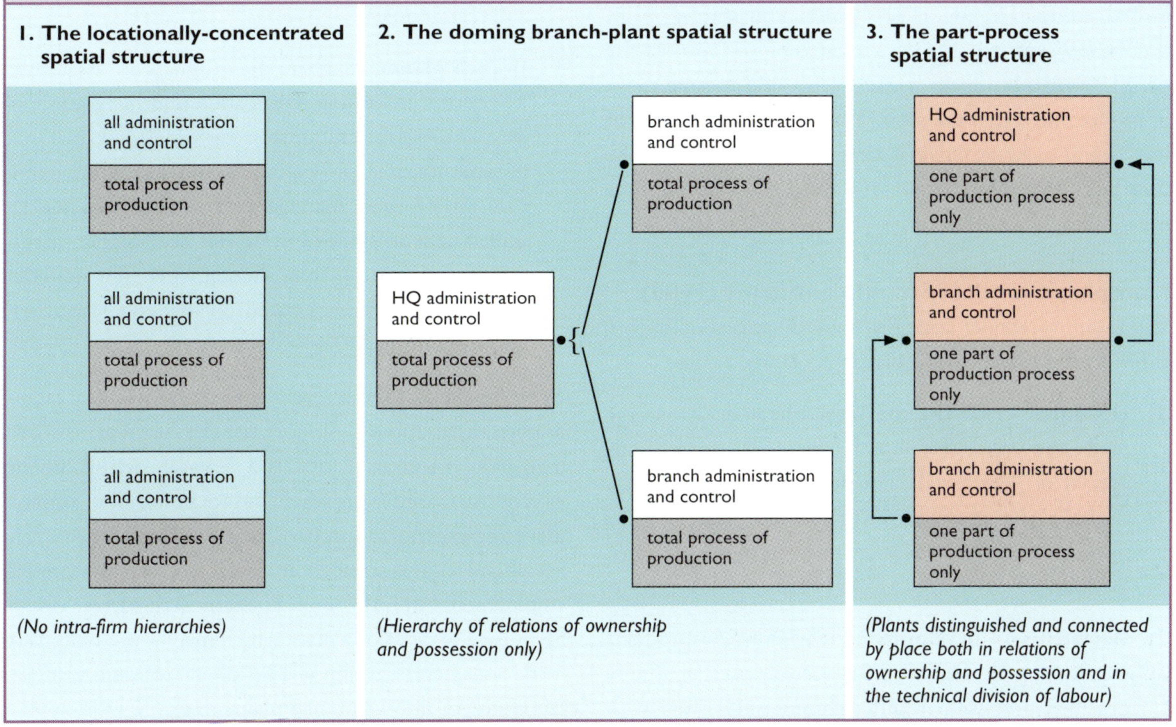

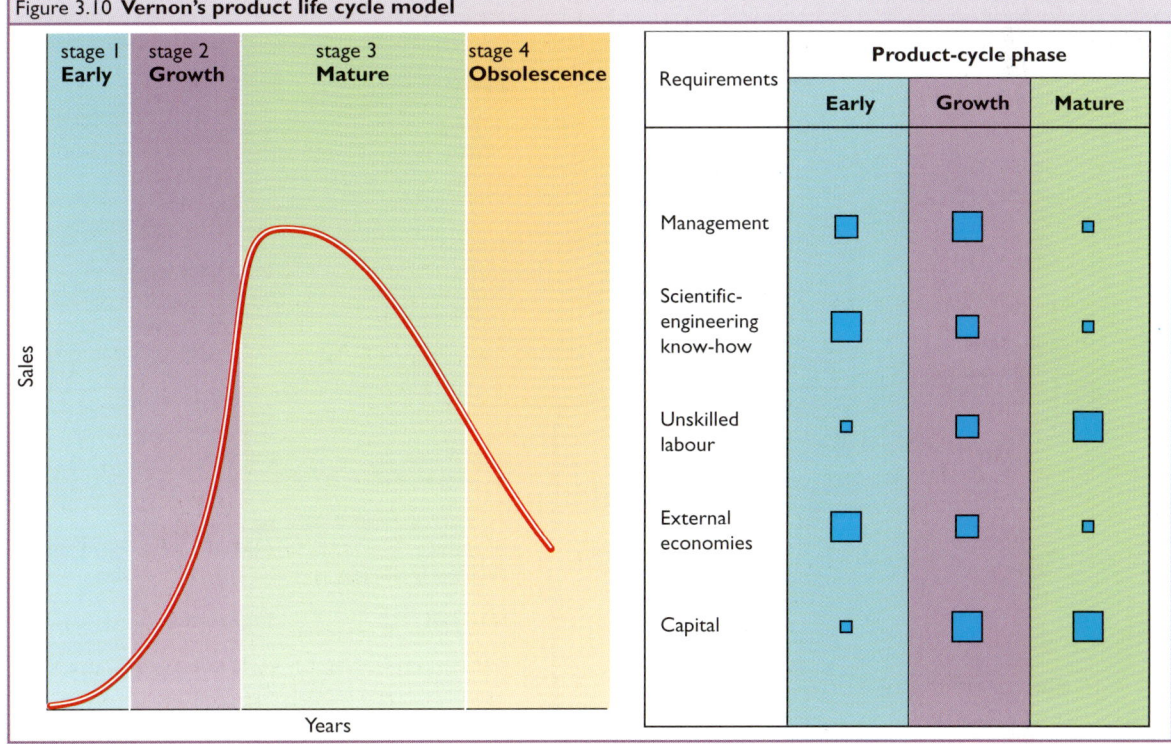

Figure 3.10 **Vernon's product life cycle model**

In many ways the product life cycle model is a model of **globalisation**. The model sees production shifting from innovation centres in the core to areas of low costs in the periphery. In a European context this could mean a move of production out of the most favoured regions to areas of low labour costs or regional assistance. There is also the possibility that production could move out of the EU to the developing world.

Humphrys' model of multiplant firms (1988)

In Humphys' model large firms are described as having three organisational levels (**Figure 3.11**):

1. **Headquarters** is the top level where decisions are made and policy developed. The workforce is managerial or secretarial. The location is usually the capital cities of the core regions although there is some evidence that HQs are moving to out-of-town locations.

2. **Research and development** is where new products are developed. The workforce comprises of highly qualified professional and technical staff often with second degrees. The location is also in the core regions but increasingly in semi-rural location in science parks and technopoles.

3. **Branch plants** are where the products are assembled. The terms processing, fabrication and integration are used to describe the process by which raw materials become finished products. The workforce are operators rather than managers or scientists. They are skilled or semi-skilled but do not need higher qualifications. Location may be coastal where raw materials are bulky. Location is based on low cost labour and government incentives.

Again, at the European level the spatial division of labour implied in this model is important. The core regions attract high-paid managerial jobs and research establishments which foster conditions which ensure innovation, employment and growth within particular European regions. The peripheral regions attract branch plants which may provide employment but little more. The term **screwdriver assembly** is used to describe a plant which merely assembles parts made elsewhere. The aim of a peripheral region is to attract reinvestment into its branch plants to secure more sophisticated and higher value-added processing activities.

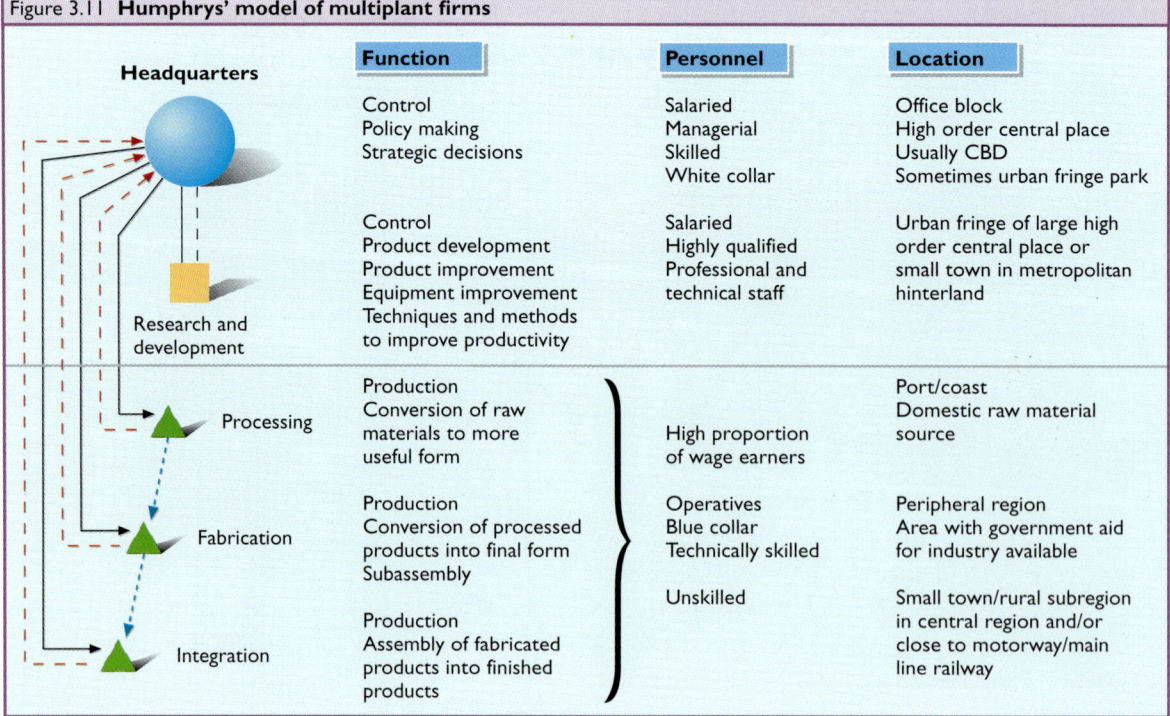

Figure 3.11 **Humphrys' model of multiplant firms**

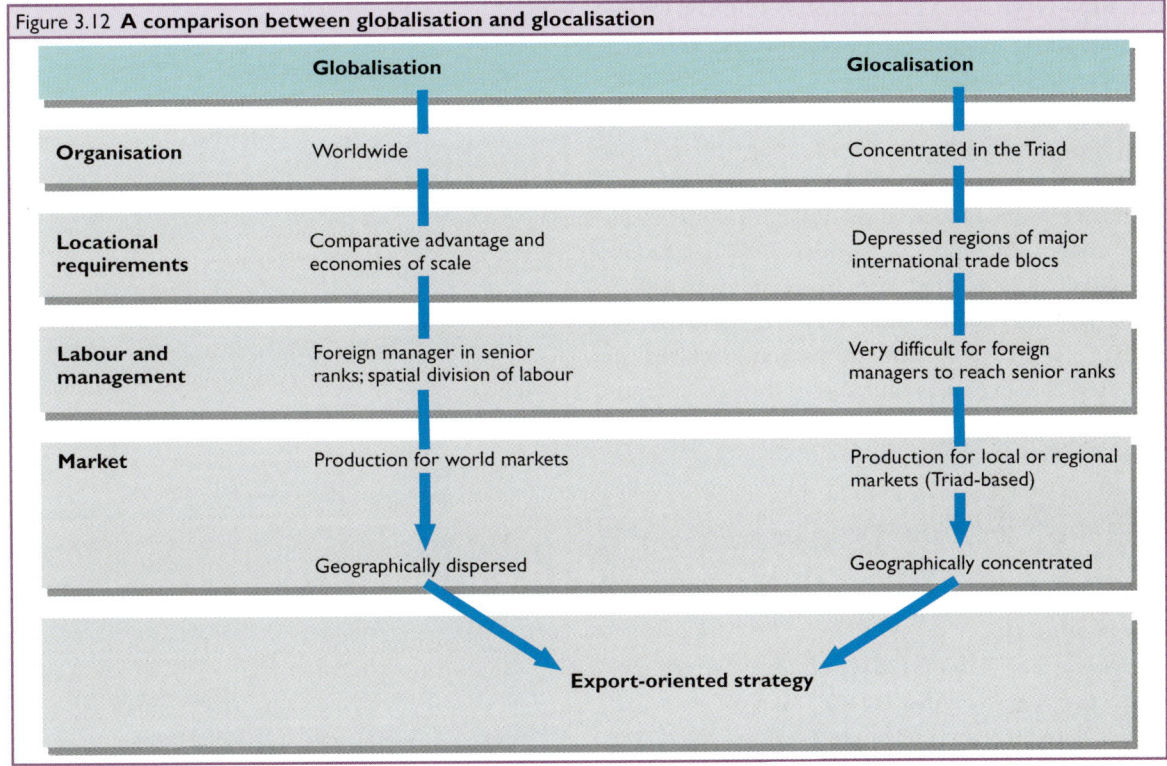

Figure 3.12 **A comparison between globalisation and glocalisation**

GLOBALISATION AND GLOCALISATION

Globalisation aims at a **worldwide** intra-firm division of labour. In this strategy, activities are established in **many sites** spread over the world, based on a country's comparative advantages. A manufacturer striving for globalisation aims to secure the supply of its inputs by locating production of these inputs at the **most favourable locations**. Thus, labour-intensive production of components will be situated in low wage areas, while the production of high-tech and high value-added parts will require a skilled or well-educated workforce. In a European context this would mean locating research facilities in core areas and assembly plants in peripheral areas. A globalisation strategy will promote a **spatial division of labour**.

Glocalisation aims to establish a **geographically concentrated** inter-firm division of labour in the three main trading blocs: Japan and South East Asia, the USA and the EU (the Triad). Manufacturers striving for glocalisation are building their comparative advantage on **close interaction** with suppliers and dealers, as well as with other relevant actors, such as banks and governments. Two essential elements stand out in a firm's glocalisation strategy:

1 The **decentralisation** of production to hierarchical networks of local subcontracting.
2 A **high degree of control** over supply and distribution.

The strategy of glocalisation involves the attempt of a manufacturer to become accepted as a 'local citizen' in a different trade bloc while transferring as little control as possible over its strategic activities. Glocalisation is first of all a political, and only in the second place a business location strategy. A manufacturer aiming for glocalisation will only localise activities in a different trade bloc if:

- it otherwise risks being treated as an 'outsider' and so subject to trade or investment barriers and thus stands to lose market share, or

- the inevitable compromise in costs and control still allow it to produce competitively, i.e. there are suitable areas of low labour costs or regional assistance.

The policy of globalisation has been associated with the US car manufacturers General Motors and their search for a world car. Glocalisation has been associated with Japanese producers, especially Toyota (**see pp. 74-6**).

JUST-IN-TIME AND FLEXIBLE PRODUCTION

Two models of manufacturing systems have been suggested. The **Fordist** model is cost driven. Here attention is paid to increasing productivity levels, lowering costs and reducing human intervention. Labour was replaced by machines but at the expense of increased specialisation and reduced flexibility. This model represents mass production. The **Toyotist** model is customer or market driven, where use of Computer-Aided-Manufacturing (CAM) and Computer-Aided-Design (CAD) has drastically reduced the time required for planning, design and production of a given product. On the shop floor the flows, inventories and costs of materials can be monitored and the use of more sophisticated robotics means that production can be geared to the market. This model is thus called flexible or lean production.

Just-in-time (JIT)

The essence of JIT is a reduction of stored parts by providing the parts when they are needed to go into a 'parent item' by the delivery of parts on same day or even every hour. Early attempts at JIT merely transferred the burden of storing parts from the major manufacturers to their suppliers. What effectively should happen with JIT is 'synthesised manufacturing', where JIT extends all the way along the supply chain ensuring each level carries lower stock.

The use of JIT has a number of key requirements:

1. **Geographic concentration.** Short distances between suppliers and consumers (assembly plants) are necessary for rapid delivery.

2. **Zero defects.** Because there is no time to reorder, all parts must fulfil stringent quality demands.

3. **Supplier network.** The number of suppliers should be small and they should operate under long-term contracts.

4. **Shared research.** There should be a transfer of technology and a research collaboration between production plants and their suppliers.

5. **Dependable workforce.** Industrial action which may delay or prevent delivery must be avoided.

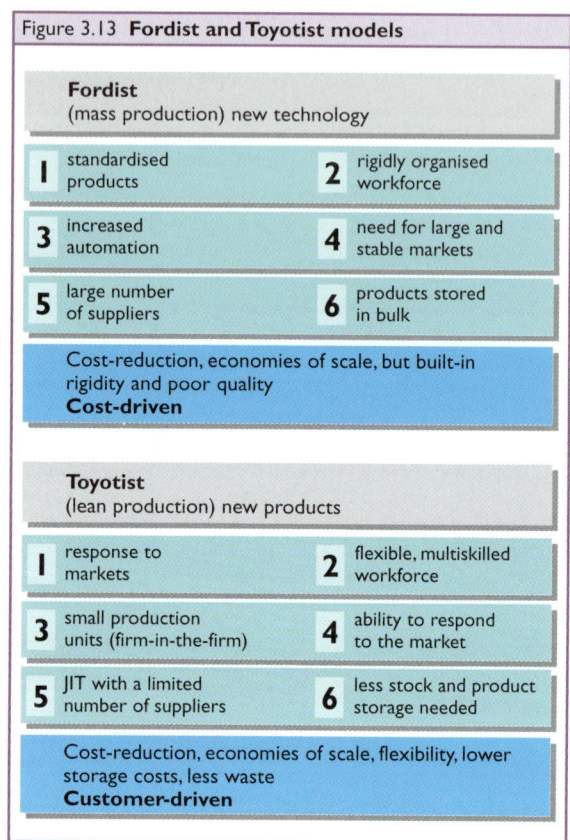

Figure 3.13 **Fordist and Toyotist models**

JIT and agglomeration

Industrial location models such as Weber's least cost model, which was discussed earlier in the chapter, stress the importance of raw materials in explaining industrial location. JIT suggests that suppliers should locate close to assembly plants to ensure rapid delivery of parts leading to an **agglomeration** of suppliers. For example, 37 new German supplier factories were built near assembly plants in order to facilitate JIT manufacturing. One instance comes from **Spain's SEAT company**, a subsidiary of Volkswagen. Volkswagen has produced the Polo model at both its home base in Wolfsburg, Lower Saxony, and at the SEAT factory at Pamplona in Spain. However, in the early 1990s Volkswagen planned to shift all Polo production to Pamplona. Spanish officials visited Wolfsburg to meet with a group of German supplier companies and persuaded three new supplier factories to relocate to Navarra. Among the attractions was to be a new German language college, for the benefit of a growing Germany community and Spanish employees alike (see pp. 78-9).

INDUSTRY AND THE ENVIRONMENT

During past two decades environmental issues have become an increasingly important part of industrial decision-making. Within the population of the member countries of the EU, there is considerable concern about environmental problems. This is true not only for the rich but also for the poor countries. In most of the countries a majority of the population also accepts that environmental protection is a necessary precondition for economic development.

Environmental technology

Environment technologies can be broadly defined within five categories:

1. **End-of-pipe equipment** installation to remove pollutants from waste streams.
2. **Waste minimisation** techniques to reduce waste per unit of input by using less polluting materials.
3. **Clean technologies** using alternative production techniques which produce less pollution.
4. **Waste management** techniques to handle, treat and dispose of waste by the most environmentally acceptable means.
5. **Recycling/resource recovery** minimising waste by reusing waste streams.

However, the use of these techniques are expensive and may adversely affect the competitiveness of EU industries. **Figure 3.14** gives a matrix of sectors relating environmental sensitivity to industrial sector. It can be seen that the sectors most sensitive to environmental pressures are either in the medium growth or low growth categories. These industries are often more highly concentrated in the peripheral or declining industrial regions of the EU.

The three scenarios

All environmental action calls for some degree of cost-benefit analysis. The level of environmental costs to industry is in the order of 1% to 2% of sales: a margin which is likely to be significant only when industries face particular economic difficulties. However, there are further costs involved in the reduced economic activity whilst industry adjusts to higher environmental standards.

Three possible scenarios have been suggested for the EU's environmental future:

1. **Green backlash**, where environmental concerns become less important and job creation, efficiency and 'value for money' are viewed as the main priorities.

2. Efficient **status quo**, where environmental pressures continue to build but companies react in traditional ways—seeking efficiencies, reducing waste, and so on, but not changing methods of production.

3. Rise of **green values**, where scientific, public and political opinion demand greater progress. Side effects of environmental degradation and a view of future problems leads to calls for a significant change in the status quo.

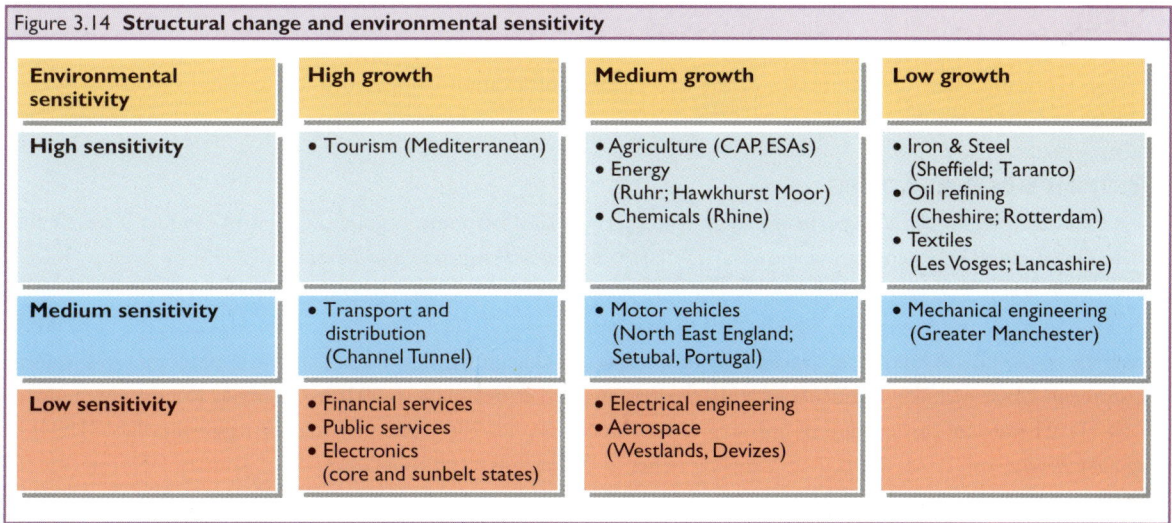

Figure 3.14 Structural change and environmental sensitivity

Environmental sensitivity	High growth	Medium growth	Low growth
High sensitivity	• Tourism (Mediterranean)	• Agriculture (CAP, ESAs) • Energy (Ruhr; Hawkhurst Moor) • Chemicals (Rhine)	• Iron & Steel (Sheffield; Taranto) • Oil refining (Cheshire; Rotterdam) • Textiles (Les Vosges; Lancashire)
Medium sensitivity	• Transport and distribution (Channel Tunnel)	• Motor vehicles (North East England; Setubal, Portugal)	• Mechanical engineering (Greater Manchester)
Low sensitivity	• Financial services • Public services • Electronics (core and sunbelt states)	• Electrical engineering • Aerospace (Westlands, Devizes)	

Section D Exercises and recommended reading

EXERCISES

1 **Figure 3.15** shows a model of changing locations in a hypothetical multiplant enterprise. Activity locations refer to the location of each product, or each stage of production which it is flexible to locate separately. The scenarios shown in B, C and D represent a process of rationalisation.

   (a) Define *rationalisation*. [4]

   (b) Describe and explain the terms *specialisation*, *concentration at an existing site* and *concentration at a new site* in the light of a firm's rationalisation. [10]

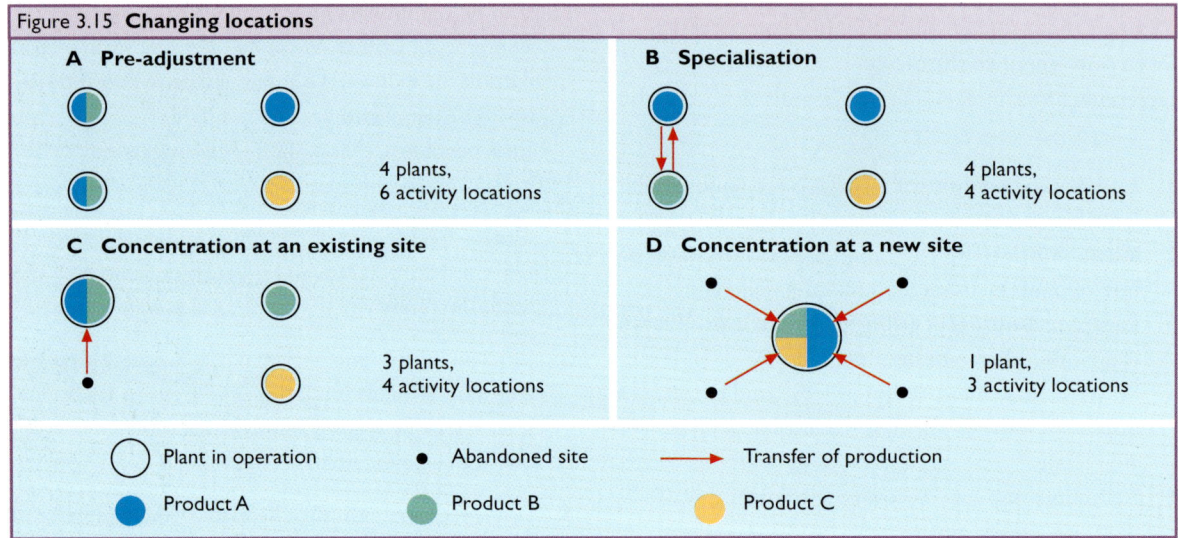

Figure 3.15 Changing locations

2 (a) What is meant by the term *spatial division of labour*. [3]

   (b) On a global scale, how would a high-tech industry, such as pharmaceuticals, be influenced by the spatial division of labour? Explain your answer. [5]

   (c) To what extent are the concepts of globalisation and glocalisation useful in explaining the spatial division of labour? [6]

3 *Essay:* Critically assess the advantages of the new theories of location compared to the classical models of Weber, Lösch and Rawstron. [25]

RECOMMENDED READING

Doreen Massey (1995) *Spatial Division of Labour* is an excellent starting point. A good overview of product life cycle theory and globalisation in general can be found in *The Globalisation of Production and Technology* (1993) by Jeremy Howells and Michelle Wood. Peet and Thrift's *New Models in Geography* Vols 1&2 (1989) provides a collection of articles by some of the leading thinkers on geographic theory. Lever's *Industrial Change in the United Kingdom* (1991) does the same but in the context of the UK. A number of articles provide examples of changing geographical patterns in economic and regional development. These include: 'Industrial location theory—in need of flexibility' (1992) in *Geography Review*, vol. 5, no. 4, pp. 38-41, 'Theories of the location of industry' (1993) in *Geofile*, no. 206, 'The new map of regional policy' (1994) in *Geography Review*, vol. 7, no. 4, pp. 7-9, 'Growing richer in the regions' (1994) in *Geographical*, January 1994, pp. 49-53, 'The East Thames Corridor' (1995) in *GeoActive*, no. 119 and 'UK manufacturing trends' (1996) in *Geofile*, no. 275.

CHAPTER 4
THE MATURE INDUSTRIES

A **COAL MINING**
  Coal mining in the UK **61**
  Coal mining in Germany **63**

B **IRON AND STEEL**
  Steel and the EU **64**
  The future **67**

C **TEXTILES**
  Changes in the European textile industry **68**
  The textile industry in the 1990s **69**

D **SHIPBUILDING**
  Swan Hunter Shipbuilders: decline in the North East of England **70**

E **VEHICLES**
  Ford and the 'world car' **74**
  The Japanese and glocalisation **75**
  The European response **76**

F EXERCISES AND RECOMMENDED READING **80**

Section A looks at the coal mining industry. This is normally classified as a primary industry, alongside agriculture and forestry, but is considered here as a mature industry on account of its close links with traditional, large-scale heavy industry. As illustrated in **Chapter 3**, coal is a heavy, gross raw material and creates a large pull on industrial development, owing to the high cost of transport. Consequently, industry has located in coalfield areas such as South Wales, the Ruhr and the Sambre-Meuse. However, many of these areas have declined as the coal industry has restructured and rationalised over the last few decades.

Section B explores the evolution of the iron and steel industry from its charcoal and coalfield sites to its modern concentration at coastal sites. Like the coal industry it too has rationalised and streamlined. It illustrates clearly many geographical issues such as deindustrialisation, environmental dereliction, globalisation and privatisation.

Sections C and D outline two traditional industries, textiles and shipbuilding, which have experienced tremendous job losses and contracting markets. The example of Northern Ireland illustrates that there are successful strategies in these industries and these are compared with losses in other regions.

Section E considers the revamped vehicle industry. Europe's motor industry still ranks alongside that of Japan and the USA, but is increasingly influenced by global forces such as multinational control, trading blocs and government policies.

Section A Coal mining

In the nineteenth and early twentieth centuries coal was the basis of industrial development throughout the world. However, since World War II it has been replaced as Europe's main energy source, although it remains important in terms of regional development. Cheap, reliable supplies allowed regional economies to expand and it was not until the increase in coal prices during the 1950s and 1960s that many countries began to import oil from North Africa and the Middle East on a large scale. This coincided with the exhaustion of many coalfields and the change in energy supplies had an impact upon the location of many industries. Most of Europe's main coalfields, including the Ruhr, Sambre-Meuse, and Nord-Pas-de-Calais (Figure 4.1), have followed a similar pattern of development: early growth and expansion followed by rapid collapse and accompanying social and economic distress. In general, there was a movement of industrial activity away from the coalfields to coastal sites. Areas which depended upon supplies of coal such as South Wales, North East England and the Ruhr were faced with the collapse of their economies and have undergone industrial readjustment, diversification and regeneration. However, the oil crises of 1973 and 1980 caused many countries to re-evaluate their energy plans and to develop policies for other supplies. Consequently, there has been reduced imports, diversification of the energy base, development of internal energy sources, increasing energy efficiency and a coordinated policy throughout Europe. These changes have had an effect upon industrial location and regional development: industries and regions dependent on large-scale energy resources have declined, whereas high-technology and high value industries have emerged and are less dependent for their location upon raw materials. The example of coal mining in the UK and the Ruhr illustrate these trends and the options for coalmining in the twenty-first century.

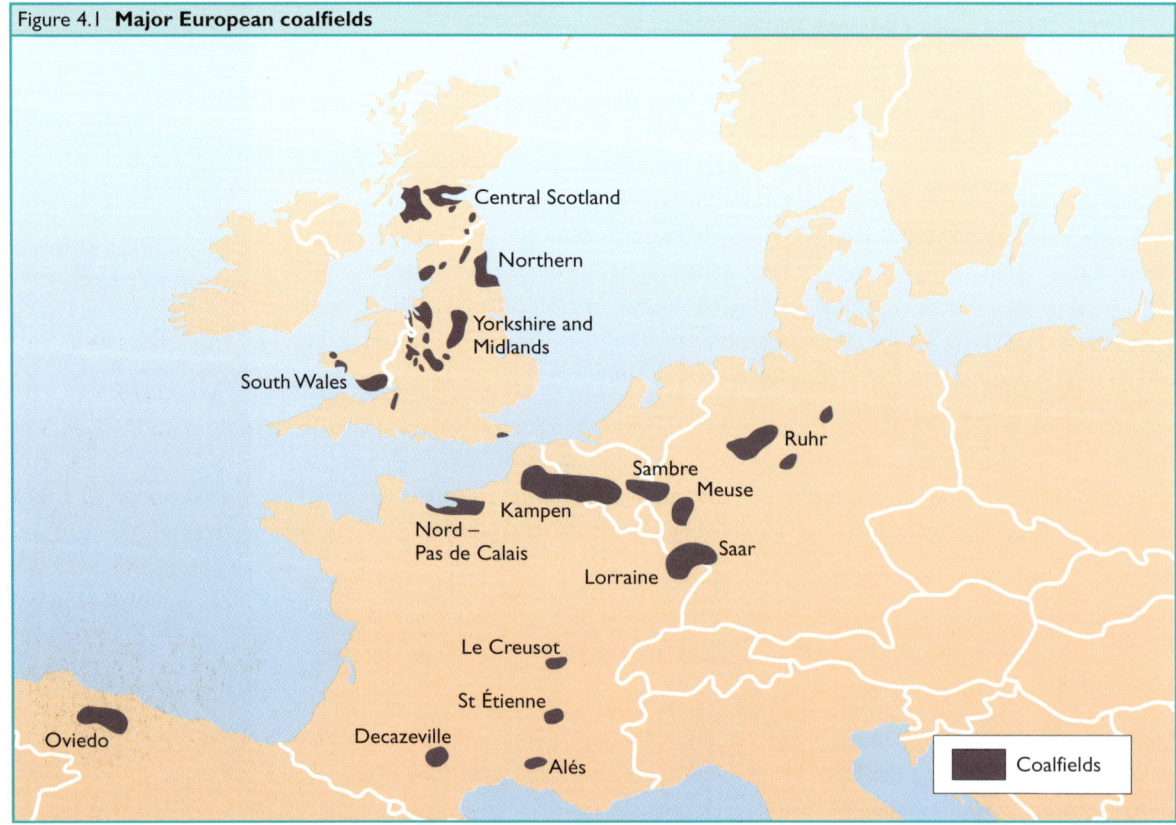

Figure 4.1 **Major European coalfields**

COAL MINING IN THE UK

Britain's coalfields are spread throughout the country, consequently industrialisation was widespread. Regional specialisation also characterised Britain's economy, for example textiles in Lancashire, steel and tin plate in South Wales, and shipbuilding and heavy engineering in Clydeside and the North East of England. Expansion occurred throughout the nineteenth century and output peaked in 1913. Since then output and employment have declined, especially since 1945. In 1947, at the time of nationalisation, Britain's coal industry employed over 725,000 miners in over 900 collieries. By the end of 1994 when British Coal was privatised less than 19,000 miners were employed in fewer than twenty collieries.

The changing fortunes of Britain's coal mining industry can be explained by a number of contrasting factors:

1. The **demand for coal has declined** dramatically since 1945. Moreover, its use has changed: domestic and industrial uses have been replaced by power stations. Moreover, during 1992-4 demand from the power stations fell by about one-third.
2. **Improvements in technology and mechanisation** have led to a huge reduction in the numbers employed in coal mining. Output per man-shift increased from 2.44 tonnes in 1982 to 8.76 tonnes in 1994. Consequently, a large number of old, inefficient collieries with high costs and low productivity were closed.
3. The government's decision in the 1960s to buy **cheap oil** from the Middle East reduced the demand for coal.
4. By contrast, the energy crises of 1973 and 1980 created **short-term booms** for coal but also increased the government's determination to develop North Sea oil and gas, as cheaper sources of energy than coal.
5. The National Coal Board's (NCB) *Plan for Coal* in 1974 was an ambitious plan to **modernise the industry** and make it more efficient and productive. Two major projects were developed: Selby in North Yorkshire and the Vale of Belvoir in Leicestershire produced 10 million tonnes and 7.2 million tonnes of coal annually.
6. Cheap oil prices in the 1980s and cheaper coal from the USA, South Africa and Australia made British coal expensive. **Imports** more than tripled between 1983 and 1990, rising from 4.5 million tonnes to 14.8 million tonnes.
7. The **miners strike** of 1984-5 turned the government against coal.
8. Increasing concern over **environmental issues** such as acid rain, water soil and air pollution has led to the reassessment of the value of coal mining (**Inset 4.1**).

During the 1990s the decline accelerated and the privatised industry of the mid-1990s bears little resemblance to the coal industry of the early twentieth century. In 1992 the government announced the closure of 31 pits and the loss of over 30,000 jobs. In 1993 the Coal White Paper was published suggesting:

1. A government subsidy for coal to meet world prices.
2. Pits threatened with closure would be offered to private enterprise.
3. More capital availability for coal technology, to meet EU legislation on carbon and sulphur emissions.
4. An increase in coal exports.

Ironically, the streamlined, rationalised industry may now be in a position to prosper:

- It is increasingly competitive.
- Britain needs a diversified energy mix, as nuclear power is not as attractive or as cheap as formerly believed, and gas is less than 50% efficient in electricity production.

However, not all coal mines will survive. In the UK the central coalfields have the best prospects whereas the peripheral ones, which have been developed for much longer, have bleak prospects (**Figure 4.2**). The privatisation of British Coal allowed RJB mining to take control of most of the English coalfields, Celtic Energy, the Welsh coal mines and Mining (Scotland) Ltd, the Scottish ones. RJB, Europe's largest independent coal mining company now produces about 80% of the UK's coal. It controls the new Asfordby deep mine, which started production in 1995. With reserves of up to 150 million tonnes of high quality coal, it is able to supply electricity producers in the Midlands with up to 4 million tonnes of coal each year.

INSET 4.1 HAWKHURST MOOR COLLIERY

Located on green belt land between Coventry and Solihull, British Coal planned to invest £450 million in a project developing coal mining in Hawkhurst Moor.

PROs	CONs
1 Up to 14,000 men are unemployed in the area: the colliery will create 700 jobs during construction and 1,800 permanent jobs.	1 Green belt land will be destroyed.
2 Up to 170 million tonnes of high quality, low cost coal will be produced.	2 Villages, such as Berkswell, will be destroyed.
3 Coal is safer then nuclear power and tough EU legislation will limit environmental damage.	3 Burning coal adds to the greenhouse effect.
4 Landscaping will ensure the colliery is not an eyesore and 30,000 trees will be planted.	4 Subsidence will affect residential and industrial property.
5 Heavy traffic will be restricted to certain roads.	5 Dust and noise will adversely effect agricultural production and ecosystems.

The plans for Hawkhurst Moor were rejected by Michael Heseltine, the then Environment Secretary, in February 1991, on the grounds that the environmental damage of the mine would be too great.

Figure 4.2 Contrasting fortunes in British coalfields

	CENTRAL	PERIPHERAL
Example	Asfordby, East Midlands	Sharlston, near Wakefield, West Yorkshire
Geology	Little folding or faulting, thick seams	Contorted strata, difficult to work, thin seams
Reserves	Large supplies	Mostly exhausted
Quality	Good	High
Supplies	Plentiful	Largely exhausted
Costs	Low	High
Mechanisation	Easy	Difficult
Productivity	High	Low
Market	Electricity generation	Exports, 'traditional' heavy industries
Overseas competition	Limited	Considerable
Regional economies	Diversified	Dependency on coal and related traditional industries
Unemployment	Below UK average	Well above average
Environment	Strict controls, evidence less obvious	Highly polluted: tips, slag heaps
Trends	Decline halted, future investments concentrated on these areas	Rapid and serious decline, major future investments unlikely
Prospects	Good: concentration on best reserves	Bleak: much smaller coal industry

Figure 4.3 **Industrial landscape, Essen, the Ruhr, 1890s**

COAL MINING IN GERMANY

German coal mining has experienced similar fortunes. In the Ruhr, the coal industry has been affected by declining orders from steel manufacturers, a lack of government finance, a slow down in electricity consumption and cheap foreign imports, only one-third the price of Ruhr coal. The decline is as sharp as in the UK. In 1970 nearly 195,000 people were employed in mining in the Ruhr. During the 1970s 50,000 jobs were lost and in the 1980s a further 40,000. By the mid-1990s fewer than 75,000 jobs remained in coal mining. In addition companies which provided mining equipment went out of business. Ruhrkohle, which still produces over 80% of Germany's hard coal, has reduced its dependence on coal since the 1970s. Increasingly it has diversified into environmental technology, engineering services and waste management. However, although it derives much revenue from these sources, up to 35% of its annual turnover, these do not provide much employment for the recently redundant unskilled labour force. Ruhrkohle is also developing new markets, notably eastern Germany and Russia.

Europe's coal mining industry is a classic example of **deindustrialisation**. These examples illustrate clearly industrial change, **rationalisation**, and **diversification**. The decline of the industry has serious **externalities** (repercussions) for ancillary industries, such as equipment manufacturers, as well as service industries affected by the downturn in the local and regional economy. Change in the Ruhr is analysed on **pages 128-31**.

Section B Iron and steel

The iron and steel industry is a good example of many of the issues facing industry in the EU. There is overcapacity within Europe and this has led to processes of rationalisation and restructuring throughout the EU. Some areas have suffered through this deindustrialisation but others have modernised to become highly efficient producers. There are tensions between private and publicly owned plants, particularly with respect to subsidies and overproduction. Recently the dominance of large integrated steelworks has been threatened by the rise of the more flexible mini-mills.

STEEL AND THE EU

Steel and coal were the subject of the Treaty of Paris (1952), one of the first European agreements. Five countries, Spain, Germany, France, Italy and the UK, account for three-quarters of the total production of steel in the EU (**Figure 4.5**). Since the mid-1970s the world steel industry has faced a fall in demand due to:

1 **Recession-based cuts** in the steel consuming industries.
2 **Technological developments** which mean less steel is needed.
3 **Improvements** in quality and durability.
4 **Substitution** of steel by other materials (aluminium, plastics and so on).

Figure 4.4 **Llanwern–Port Talbot integrated steelworks**

Competition

From 1980 to 1988 the EU steel industry was protected from foreign competition by the **d'Avignon Plan** which was a rough mix of **quotas, subsidies** and **protectionism**. When it expired in 1988 the industry was exposed to competition from Central and Eastern Europe, China, South Korea and Turkey. Privatised companies like British Steel welcomed the chance to rationalise and restructure but other companies, especially the then state-owned Italian and Spanish plants, have been slow to change.

The two paths

In addition to the rising competition, the EU has an overcapacity of 30 million tonnes. The options facing the European steel industry have been outlined by Mr. Maarten van Veen, Chairman of Hoogovens, the Dutch steel maker. He suggests two possible paths:

1 **Déjà vu.** This is a return to the subsidised and protectionist path encapsulated by:
    ♦ a refusal to look beyond national boundaries;
    ♦ sheltering behind trade barriers and state subsidies;
    ♦ viewing the global market as a threat;
    ♦ producing standard, 'cost-driven' products.

2 **New era.** This sees the creation of a commercially driven international, or even global steel industry:
    ♦ with a global approach;
    ♦ with a move towards privatisation;
    ♦ viewing the global market as an opportunity to specialise or expand;
    ♦ making 'dedicated' products tailored to customers needs.

Van Veen's two paths have a number of repercussions with respect to the location, restructuring and technological change in the industry.

Regional distribution

There are steel plants throughout the EU with varying levels of production capacity. Almost half of EU crude steel production comes from only five companies: Thyssen Stahl (Germany), Usinor-Sacilor (France), Ilva (Italy), British Steel (UK) and Hoesch (Germany).

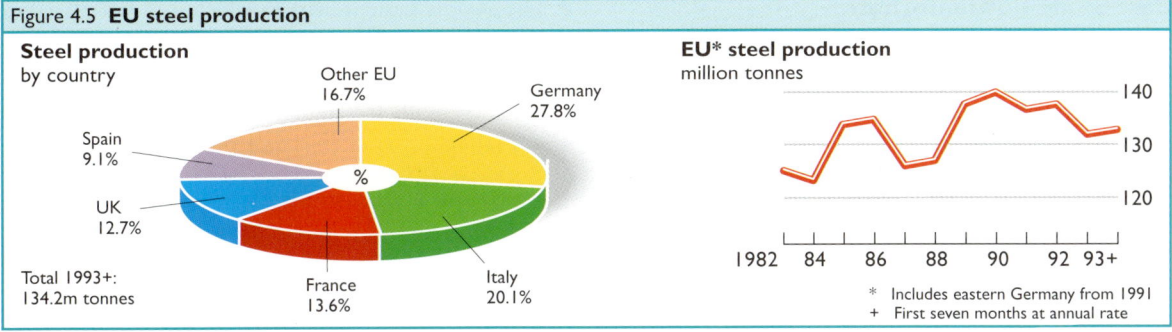

Figure 4.5 **EU steel production**

Steel plants were originally built on inland sites close to raw materials, especially coal, or near steel consumers. However, the economies of scale of **large integrated steelworks** have drawn plants to coastal locations. Such sites offer cheap, flat land but more importantly access to imported raw materials and new, international markets (**Figure 4.4**).

The steel industry can no longer be considered the classic industry of **Weberian** least cost location. In Europe, low grade orefields have been replaced by **cheap imports** from Australia and Brazil. Transport costs by sea are not related to distance. Rather it is the costs of loading, unloading and storage at port facilities which are high. **Deepwater ports** offer access to raw materials without the need for a costly change in the type of transport (break-of-bulk). Weberian analysis can still be used but only to explain closeness to a **break-of-bulk** site rather than the real source of the raw materials which may be thousands of miles away.

Restructuring and rationalisation

The European Commission has tentatively addressed overcapacity and overproduction in the steel industry for the last 20 years. During the 1980s rationalisation reduced production capacity for crude steel by 40 million tonnes (19%). Productivity (output per worker) rose by 62% between 1983 and 1992 due to a massive reduction in workforce. However, the rationalisation was only undertaken by the EU's most efficient producers in the UK and France. For example, technological development in the UK's Port Talbot plant saw the following changes:

- In the 1950s and 1960s Port Talbot was nicknamed 'Treasure Island' because of its seemingly endless supply of jobs and high wages.

- In the late 1960s it was recognised that the plant was overmanned: 16,000 workers produced 3 million tonnes/year with equivalent labour costs of £37/tonne.

- By the 1980s the workforce had been reduced by more than half and labour costs were £14/tonne.

The **Italian and Spanish** producers have been more reluctant to change. Indeed, the restructuring plan has been seen as a classic case of mismanaging the decline of a capital-intensive industry. Policy in the 1980s has been described as 'a great machine for slowing down the exit of high cost producers.' There are a number of problems associated with closing steel mills:

1. It is **cheaper to keep a plant running unprofitably than to close it completely**—in a total shutdown, the company must write-off the entire value of the factory, make heavy outlays to dismissed workers and pay to dismantle the mill and clean up the site.

2. **Subsidising inefficient producers** has seen about £60 billion in government money spent since the mid-1970s through waivers and loopholes in the legislation.

3. As France, Germany and Britain were closing mills the subsidised Spanish industry built **new plants** that boosted its capacity by 35%, while Italy raised its potential by 16%.

An unratified EU plan of 1993 aimed to address some of these problems and aimed for a reduction of capacity by 30 million tonnes at an expected cost of around 50,000 jobs, with EU money available to help with costs of redundancies.

The rationalisation and restructuring introduced by the Italian producer Ilva and British Steel illustrates some of the problems and opportunities which the EU faces.

Case Study 1: Taranto, Italy

The steelworks owned by Ilva at Taranto, located in the heel of Italy's Mezzogiorno, has long been used as an example of a growth pole strategy. It is the biggest steelworks in Europe, capable of producing 10 million tonnes of crude steel and occupying a site of over 1,500 hectares. After 30 years as a nationalised and subsidised plant it has finally been privatised.

The plant has long been the target for criticism from unsubsidised German and British steel makers, although the Italians claim that they have restructured and the plant is viable. The plant has a number of advantages:

1. It has a modern site next to a deep-sea port for raw material deliveries.
2. It has closed down two reheating furnaces, cutting capacity by 1.2 million tonnes a year and reducing the workforce by 2,500.
3. The town of Taranto benefits from LIRE 1,100 billion (£440 million) in wages and services.
4. The costs of production fell by 14% between 1990 and 1993.

And yet even after privatisation the plant faces problems:

- It has not acted as the growth pole for the Mezzogiorno as was hoped.
- The Middle Eastern markets which it is well placed to serve have not materialised and it looks poorly placed to tap other international markets.
- Subsidies to the Italian industry cost tax payers close to £14 billion between 1981 and 1992.

Case Study 2: British Steel

British Steel has been one of the fiercest critics of subsidised state-owned plants like Ilva at Taranto. **Figure 4.6** shows the extent of the company's rationalisation and restructuring. It claims, with some validity, to be one of the world's most efficient and competitive steel companies:

1. It is Britain's seventh largest manufacturing exporter with sales of over £578 million in 1994-5.
2. It cut its workforce from 166,000 in 1980 to 36,000 in 1995 but now produces more steel.
3. It reduced its fixed costs by £250 million between 1992 and 1995.

However, the increase in productivity has been achieved at considerable social costs. The closure of plants has seen a rise in unemployment in some areas (**negative deindustrialisation**) but also improvements in the competitiveness of remaining plants and new opportunities (**positive deindustrialisation**).

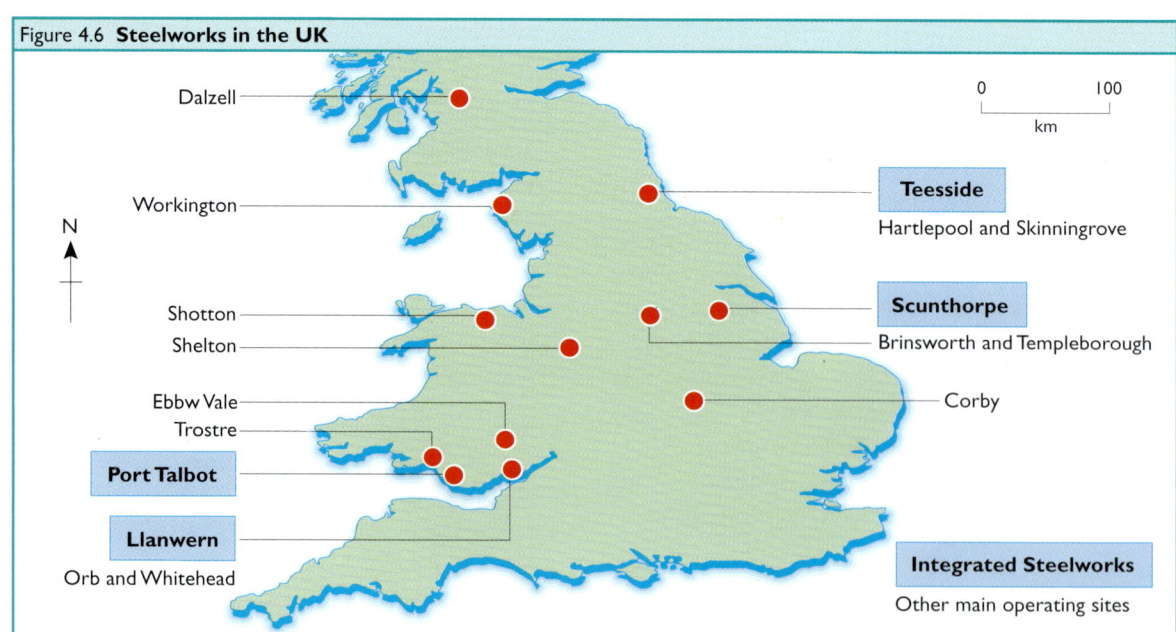

Figure 4.6 Steelworks in the UK

Negative deindustrialisation: Ravenscraig

The closure of the steelworks at Ravenscraig in Scotland in 1992 illustrates many issues resulting from deindustrialisation. The privatisation of British Steel has been blamed for the closure but in reality it merely acted as a catalyst. The truth is that Ravenscraig was kept alive due to industrial inertia and failed as a regional growth pole for a Scotland growing in high-tech, high value-added industries. Four factors lie behind the closure:

1. British Steel could save £70 million a year in overheads by **concentrating production** at two sites: the Llanwern–Port Talbot Complex in south Wales and Teesside in the North East.
2. **Recession and decline** in steel markets—at the end of 1992 galvanised steel was selling for £260/tonne, a decline of 33% from 1989.
3. A change from a national to a **European focus**: Ravenscraig was kept running to stimulate Scottish manufacturing but only 4% of its steel went to Scotland.
4. British Steel is now a **private-sector company** with an eye on profits and its subsidised competitors, rather than the Scottish electorate.

Positive deindustrialisation: Sheffield

In 1942, 176 furnaces produced 1.92 million tonnes of steel in Sheffield. In 1992, seven furnaces made 2.25 million tonnes of top grade steel. The Lower Don Valley, the centre of manufacturing for decades, has seen a resurgence from a period between 1974 and 1986 when eight large steel companies closed and unemployment peaked at 18% in 1984. A number of strategies have been introduced to reindustralise the city:

1. Links between industry and the city's two universities have been improved to encourage **technology transfer and research**.
2. The **city centre is being redeveloped** with a £250 million investment in the South Yorkshire supertram.
3. **Derelict land is being reclaimed**: by 1997 650 acres will have been reclaimed giving 6.3 million square feet of industrial and commercial space and creating 18,000 jobs.

Sheffield's rationalised steel industry and diversified commercial base shows that deindustrialisation is often necessary if a region or industry is to advance.

The future

The future of steel production is still poised between Van Veen's two paths. If the steel industry is to rationalise its European operations further it must use two strategies:

1. Separate the commercial issue of capacity cuts from the political one of employment by investing in retraining schemes for redundant workers.
2. Prevent existing steel makers from using capacity restrictions as barriers to keep out new, low-cost rivals, particularly mini-mills (**Inset 4.2**)

INSET 4.2 **MINI-MILLS**

There are two major categories of steel plant:

1. *Integrated steelworks, which produce steel from iron ore in blast-furnaces and cast it into slabs, plates, coils or sections. They have a capacity of 2–10 million tonnes.*

2. *Mini-mills are more specialised mills which turn scrap into steel by melting it in electric arc furnaces. They have a capacity of 0.15–2.5 million tonnes.*

Mini-mills developed in the 1960s and although integrated steelworks still account for 70% of steel production in the EU the significance of mini-mills is rising. They account for 55% of production in Italy and 42% in Spain. The debate for and against mini-mills is like the argument for lean production against mass production in the car industry.

FOR MINI-MILLS	AGAINST INTEGRATED
Lean and energy efficient	Economies of scale but only at full capacity
Environmentally friendly recycling, so that overproduction is encouraged	Polluting
Entrepreneurially managed	Often nationalised
Customer-driven	Cost-driven

Section C Textiles

CHANGES IN THE EUROPEAN TEXTILE INDUSTRY

A significant change in the locational importance of factors affecting the textile industry is apparent. In the nineteenth century proximity to water and coal supplies, cheap labour supplies (often female) associated with traditional heavy industry areas, access to coastal transport facilities, and effectively a monopoly on the European market led to the development of major textile areas, such as Lancashire-Yorkshire, northern Italy and northern France-Flanders-Ruhr. A number of secondary areas also developed: the Rhone Valley, Iberia and Northern Ireland.

By contrast, in the late twentieth century electricity supplies give a greater locational flexibility: most raw materials are imported from developing countries and machinery has reduced the need for skilled labour, although **inertia, specialisation** in high quality goods, and lower labour costs in peripheral areas maintain the importance of the textile industry. As the market and the manufacturers have become increasingly global, Europe's textile producers have fallen behind many developing countries.

The main trends in the textile and clothing industries can be summarised as:

1 In developed countries, **major shifts in demand** away from standardised, mass produced items to fashion and quality. This retail revolution necessitated shorter production runs and the need for a quick response to fashion changes. For example, in the mid-1990s Marks and Spencer cut scores of suppliers and concentrated on those that could deliver high quality goods rapidly.

2 Intense **international competition from Asia.**

3 **Lack of competitive edge** caused by years of neglect and under-investment.

4 **Relocation of multinational plants** overseas to benefit from lower labour costs.

5 **Organisational disintegration**, making branch plants legally separate enterprises, although financially linked to the parent company: smaller enterprises have lower wage costs and are more responsive to consumer demand.

6 **More specialised and segmented markets**, better served by smaller enterprises. Disintegration thus offered an alternative to branch plant closure for large firms to maintain their profit levels.

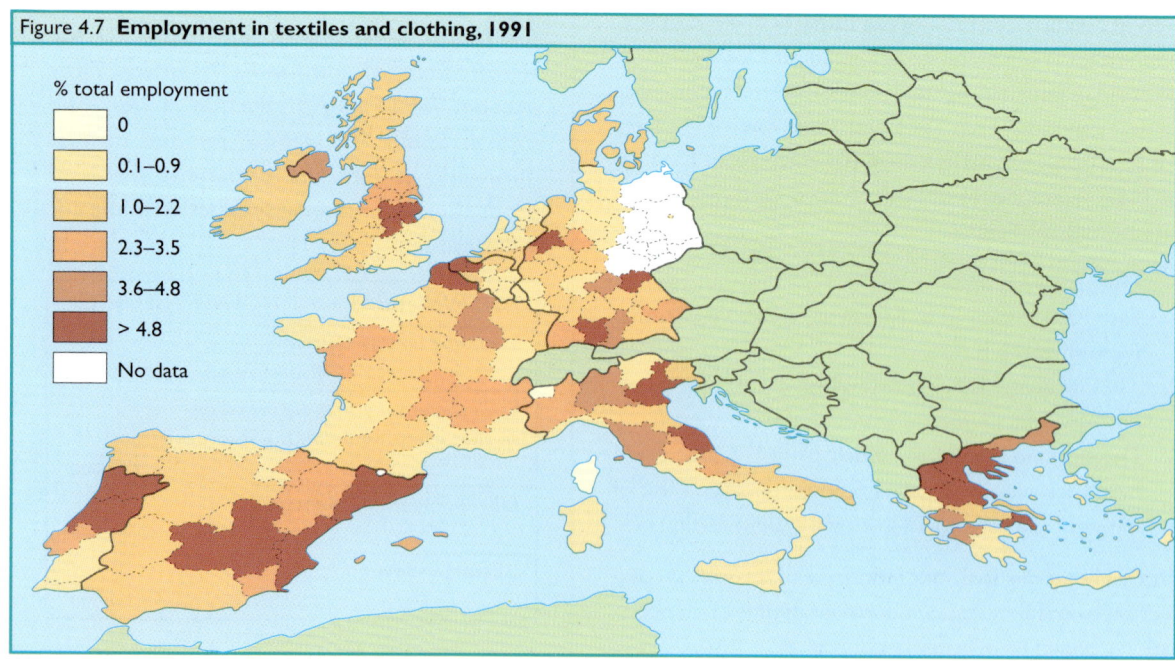

Figure 4.7 **Employment in textiles and clothing, 1991**

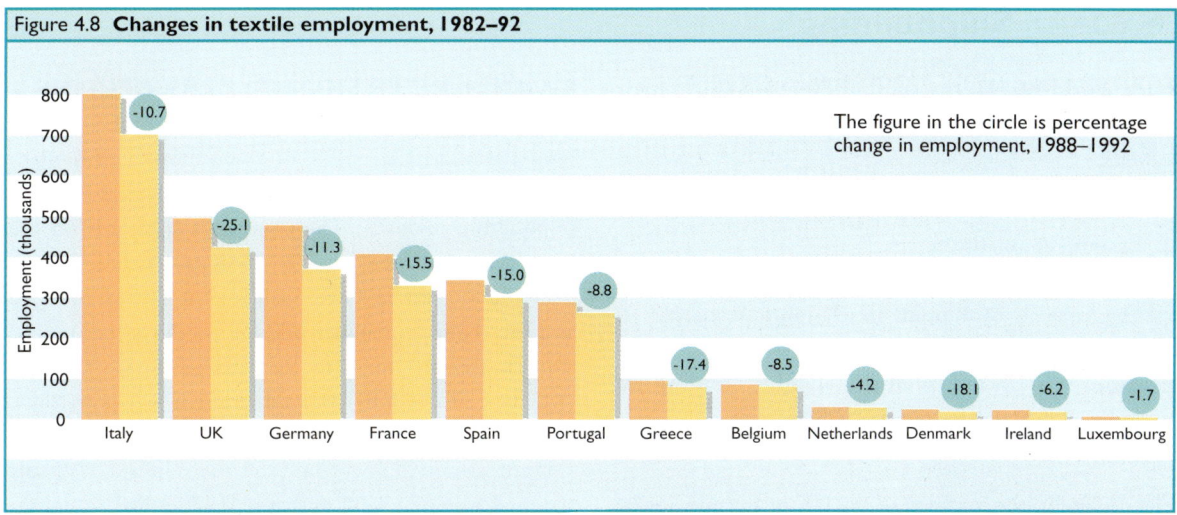

Figure 4.8 **Changes in textile employment, 1982–92**

The figure in the circle is percentage change in employment, 1988–1992

Textiles in Les Vosges, France

The French textile industry is long-established, centred on old industrial areas such as **Nord**, **Rhone-Alpes** and **Alsace-Lorraine**. Although there has been a significant decline in the industry, and many closures, it is still important. The Alsace-Lorraine region is the main cotton textile region. The cotton industry of Alsace was more technologically developed than other areas because of its association with other industries, which provided the local manufacture of machinery and advanced chemical dyes.

Epinal, capital of the **Les Vosges** department, was the centre of the region's cotton manufacturing and its chemical industry was the basis of an important synthetic fibre industry. **Epinal** is an excellent example of how to respond to the long-term decline of a traditional industry. Until the 1960s, it was the thriving capital of the French textile industry, known for its *Images d'Epinal*, popular cotton prints of French historical scenes. Since then 35,000 textile industry jobs have been lost in the area owing to low-cost Far Eastern competition. By the mid-1990s textiles employed only about a quarter of the workforce. Epinal survived the downturn in textiles surprisingly well, by **retraining** its skilled yet cheap workforce to other sectors, especially paper, another traditional Vosges industry, and engineering, which between them now employ a quarter of the workforce. Other factors have been important in this **restructuring**. The surrounding **softwood forests** and proximity of **cheap nuclear electricity** have attracted Scandinavian paper companies searching for an entry to continental Europe.

THE TEXTILE INDUSTRY IN THE 1990s

Although the European clothing and textile industry has undergone a tremendous decline in terms of employment it still employs nearly 3 million people. However, estimates for employment in the year 2000 range from 0.75 million to 1.5 million with most jobs lost in the low wage countries, such as Portugal.

Although turnover in textiles and clothing rose slightly in the early 1990s, consumption, production, employment and investment fell quite considerably. The fall in employment in textiles was exceeded only by steel in relative terms and by manufacturing industry in absolute numbers. Moreover, the decline has not been even throughout Europe. The largest decreases have occurred in the UK, Denmark and Greece whereas Germany and Italy have been less affected. In the UK the decline occurred early: between the 1960s and the mid-1980s employment in garment manufacture fell from 800,000 to 246,000, and then to 170,000 in 1995.

The future of the European textile industry now seems to focus on two areas:

1. Production of high quality, high value goods: skills and tradition enable this to occur in long-established textile areas, e.g. Flanders, Lancashire–Yorkshire and northern Italy.

2. The lower labour costs in the Mediterranean area will favour Spain, Portugal and southern Italy.

Section D Shipbuilding

Since World War II, the shipbuilding and ship-repair industries in Europe have declined for a variety of reasons:

1. A decrease in orders due to overseas, mainly Far Eastern, competition.
2. World recession and rationalisation of shipyards.
3. Overcapacity of shipping in relation to demand.
4. Competition from other forms of transport.
5. More widespread political stability leading to reduced demand for naval warships.
6. Lack of investment in the industry.

The decline is not confined to the UK or Europe: in the early 1990s Far East producers have been particularly badly hit. In Japan, falling demand for new ships and workload for its yards was keenly felt. The continued production of some British shipbuilders is all the more remarkable given the high level of subsidies offered by many countries. In the USA, for example, government finance is available for up to 87.5% of the cost, repayable over twenty-five years compared with between 30% and 60% in the UK, repayable over eight and a half years. The two case studies that follow illustrate the changing fortunes of the shipbuilding industry, and, in particular, the effect of decline in the defence industry.

SWAN HUNTER SHIPBUILDERS: DECLINE IN THE NORTH EAST OF ENGLAND

From the mid-nineteenth century until the late twentieth century, Swan Hunter played a leading role in the British shipbuilding industry. The company's shipyards on the River Tyne constructed over 2,700 ships—including more than 400 warships and fleet auxiliaries. It was first founded as a merchant shipbuilder offering products that ranged from passenger liners to Very Large Crude Carriers (VLCCs) although by the mid-1990s it was primarily a builder of warships for the Royal Navy.

From its development in 1860 the shipyard gradually acquired adjoining land so that by 1930 it had an unbroken river frontage of some 4,000 feet and works covering nearly 80 acres. The company dealt with all aspects of shipbuilding, i.e. design and construction, repairs, overhauls and renewals. During the two world wars, it constructed, repaired and converted naval and merchant ships. After the war there was massive replacement of shipping lost during the war and it was evident that the layout and structure of the shipyard had to be changed if the company was to be competitive in the modern shipbuilding industry.

Following the Geddes Report on shipbuilding (1966) discussions were opened with the owners of other ship-

Figure 4.9 **Swan Hunter at its peak: the launch of the Mauretania, 1911**

Figure 4.10 **The last ship sails out of Swan Hunter, 1994**

builders on the Tyne on the possibility of merging all the shipyards on the Tyne into one company. Between 1972 and 1977 almost £16 million was spent on modernising the yards. By 1977, when Swan Hunter Shipbuilders Limited was nationalised, all the shipyards on the Tyne were part of the one company. Approximately 11,500 were employed at the yards building cargo vessels, bulk carriers, tankers and container ships of all sizes.

In 1986 Swan Hunter was re-privatised in a management buy-out. The company decided to concentrate on warships and auxiliaries for the Royal Navy. Shipbuilding was concentrated at Wallsend and employed some 3,500 people. The Neptune Yard contained a pipe factory, a module shop and general engineering works while facilities in the Hebburn Dock concentrated around the drydock. However, in face of increasing competition, a falling-off of orders from the UK MOD and a failure to win significant export orders Swan Hunter was forced into receivership in 1993 following an unsuccessful attempt for an amphibious assault helicopter carrier.

By the middle of 1994 the closure of Swan Hunter became almost certain after the MOD awarded a £40 million contract to refit the ship, *Sir Bedivere*, to its Scottish rival, Rosyth Royal Dockyard. Although the yard won outfitting contracts for Type 23 Frigates (its main workload) this was not enough to guarantee continued employment for its 2,000 workforce. A potential buy-out by the French shipping company, Constructions Mechaniques de Normandie (CMN), did not materialise. The key factor was the refusal of the MOD to provide Swan Hunter with sufficient workload, a result of the reduced demand for military craft, reduced military expenditure and the saving of the Rosyth Naval Dockyard by offering contracts there. However, in June 1995 the Swan Hunter yard was bought the THC group, specialists in offshore oil platforms. By this time only 30 people were employed on the yard and although shipbuilding has ceased in the North East of England the last remaining yard is still open and could revert to shipbuilding in the future. The decline in shipbuilding in the North East was partially offset by the announcement by Siemens, the German manufacturer, in August 1995, that it intended to locate its new microchip plant in the North East, thereby creating 2,000 jobs in a £1.13 billion project.

INSET 4.3 TEXTILES AND SHIPBUILDING IN NORTHERN IRELAND

TEXTILES

Northern Ireland has a long history of textiles and clothing manufacture, dating from the linen industry in the 1700s. The tradition and enterprise which were built up within the industry subsequently resulted in diversification into other textile products, both natural and man-made. It also generated several important ancillary crafts leading to the growth of the modern clothing industry.

Textiles and clothing currently represent 26% of all manufacturing jobs in Northern Ireland, employing around 26,000 people: 10,000 in textiles and 16,000 in clothing. The 1990 annual sales in each sector were £400 million and £480 million respectively, representing some 15% of the province's manufactured goods. The strength of Northern Ireland's textile and clothing industry reflects its strong traditional base in the sector, the growth of the British chain stores which account for about 60% of clothing sales in Britain, providing a large, reliable demand, and, thirdly, the high level of government support available, up to 40% more than on mainland Britain, since the start of the Troubles in 1969.

The industry is diverse in structure and is made up of numerous manufacturing companies, including large multinational organisations and small private enterprises. Of 388 textiles and clothing firms surveyed, over 40% employed ten or less people whereas nine companies employed over five hundred people. The industry is spread throughout the province and provides stability in many rural areas and small provincial towns where alternative employment opportunities are limited. Although the textile industry is highly competitive, Northern Ireland's manufacturers are at an advantage over their Far Eastern rivals as they can respond more rapidly to changes in demand. Far Eastern suppliers rely on cheap labour and ocean transport and to respond quickly requires expensive air freight. The province's textile industry looks set to increase in importance with the arrival of the Taiwanese group, Hualon, near Belfast, in a project that will create 1,800 new jobs, aided by over £60 million of grants from the UK government (pp. 136-8).

SHIPBUILDING: HARLAND AND WOLFF

Shipbuilding developed in Belfast owing to its favourable site and situation. The Harland and Wolff shipyard started in 1858 and two innovations greatly improved its reputation: increased length of ships, giving greater carrying capacity without any decrease in speed, and steel decks strengthening the hull. The shipyard built a number of very large ships, including the Oceanic and the ill-fated Titanic, and during the two world wars it constructed 139 naval vessels, including 6 aircraft carriers, as well as 130 merchant ships.

Peak employment occurred between 1945 and 1950 with 30,000 people employed at the yards. However, from the late 1950s problems in the shipbuilding industry intensified: air travel replaced ship travel across the Atlantic, and foreign competitors took an increasing share of the market. Consequently, Harland and Wolff concentrated on building some of the largest vessels in the world. This was facilitated by a huge building dock, over 400 metres long and nearly 100 metres wide, accommodating ships of up to 1.2 million tonnes dryweight, twice the size of the largest ship ever built.

Continued competition and global recession following the rise in oil prices rises in 1973 reduced the amount of work carried out by the shipyard. Employment fell from 12,000 in 1969 to 8,500 in 1979. To compete Harland and Wolff diversified and rationalised: labour costs were

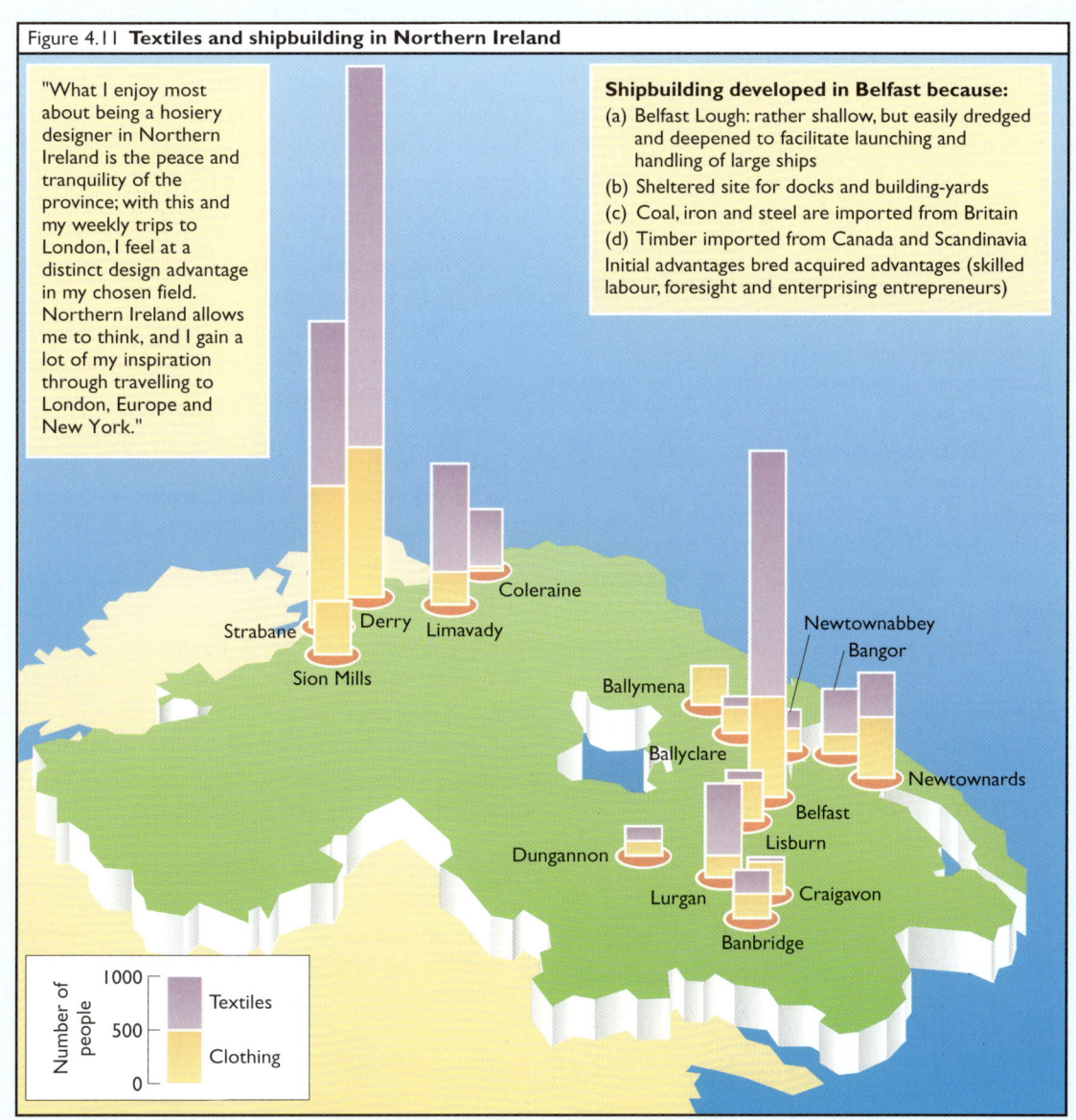

Figure 4.11 **Textiles and shipbuilding in Northern Ireland**

"What I enjoy most about being a hosiery designer in Northern Ireland is the peace and tranquility of the province; with this and my weekly trips to London, I feel at a distinct design advantage in my chosen field. Northern Ireland allows me to think, and I gain a lot of my inspiration through travelling to London, Europe and New York."

Shipbuilding developed in Belfast because:
(a) Belfast Lough: rather shallow, but easily dredged and deepened to facilitate launching and handling of large ships
(b) Sheltered site for docks and building-yards
(c) Coal, iron and steel are imported from Britain
(d) Timber imported from Canada and Scandinavia
Initial advantages bred acquired advantages (skilled labour, foresight and enterprising entrepreneurs)

reduced, flexible working hours introduced and productivity doubled. Although the shipyard now specialises in enhanced structural and safety features, it covers general cargo vessels, passenger/vehicle ferries, whaling ships, oil products carriers, bulk carriers and liquefied gas carrying ships. Moreover, the engine building yard has been transformed to general engineering. However, employment has continued to fall, 5,500 in the mid-1980s down to 2,000 or so in the mid-1990s and it needs to build between four and six new ships each year, about 5% of all global orders, if it is not to follow the fate of Swan Hunter.

Section E Vehicles

The European car industry employs more than 1.8 million people in the supply and manufacturing chain with around the same amount employed in the distribution and repair sector. The industry accounts for nearly 2% of the EU's GDP.

Figure 4.12 shows that the car industry is a global industry dominated by a triad of trading areas—the EU, the USA and Japan. In Europe, the industry faces not only the effects of a downturn in demand but also increased competition. The main pressures will come from US and Japanese companies which have set up modern efficient transplant factories in the EU. There is also competition from areas outside the Triad which have a lower cost base, such as South Korea and Malaysia. The South Korean manufacturers have seen their EU market share rise from 0.3% in 1991 to over 1% in 1995.

The car industry has also seen a change in the way cars are produced. The move from mass production ('Fordism') to lean production ('Toyotism') was discussed in Chapter 3. The US, Japanese and European producers have each reacted slightly differently both to the competitive squeeze in Europe and the new production techniques.

FORD AND THE 'WORLD CAR'

Ford and General Motors have decades of experience in Europe and share common locational patterns and production techniques. Whilst they have embraced some of the elements of **JIT production** (Inset 4.4) introduced by the Japanese they still aim to be global in their outlook. In the case of Ford this means a return to the concept of a **world car** which was introduced in the 1970s. This involves savings achieved by developing a common product which can be manufactured and sold in different countries. The research, assembly and **sourcing** is divided between locations in the USA, Mexico and Europe leading to a spatial division of labour. Decision-making is also shared. The result has been the merging of its European and North American operations to create five **vehicle programme centres** (VPCs), four in North America and one in Europe. The European VPC, with research centres in the UK and Germany, will be responsible for small and medium front wheel drive cars. It will determine policy for plants not only in Europe but also in the USA and Mexico.

Ford's 'world car', the **Mondeo**, represents another example of this merging of operations. It is a £6 billion attempt to replace the Sierra in Europe and the Ford Tempo/Mercury Topaz in the USA. The car is being sourced and assembled in a truly **global** way:

1. Assembly takes place at Genk in Belgium, Kansas City in the USA and Mexico.
2. The four-cylinder engine is being made at plants in Bridgend in the UK and at Cologne in Germany for Europe, and at Chihuahua in Mexico for North America.
3. The executive car engine is being made at Cleveland, Ohio, for both the US and European-produced cars.
4. Manual transmissions are being made in Europe at Halewood in Merseyside and at Cologne; gearboxes are made in the USA.

This strategy includes most of the elements of **globalisation** which were described in Chapter 3: spatial division of labour, production for world markets, shared decision-making and decentralisation of activities from the parent country. This is in contrast to the glocalisation adopted by some of the Japanese producers in Europe.

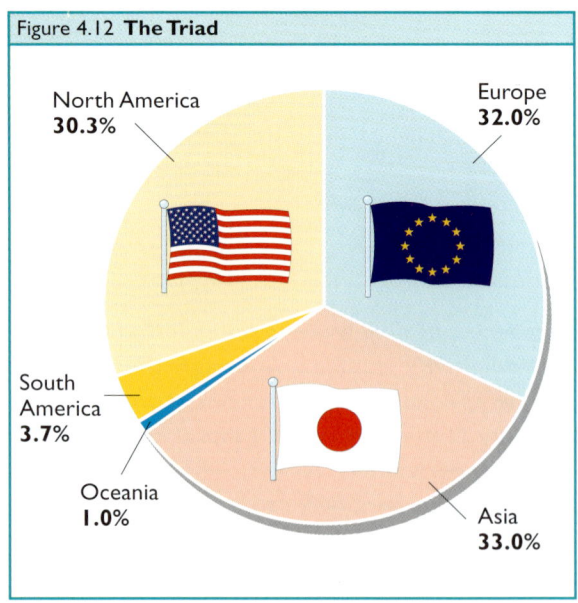

Figure 4.12 The Triad

North America 30.3%
Europe 32.0%
South America 3.7%
Oceania 1.0%
Asia 33.0%

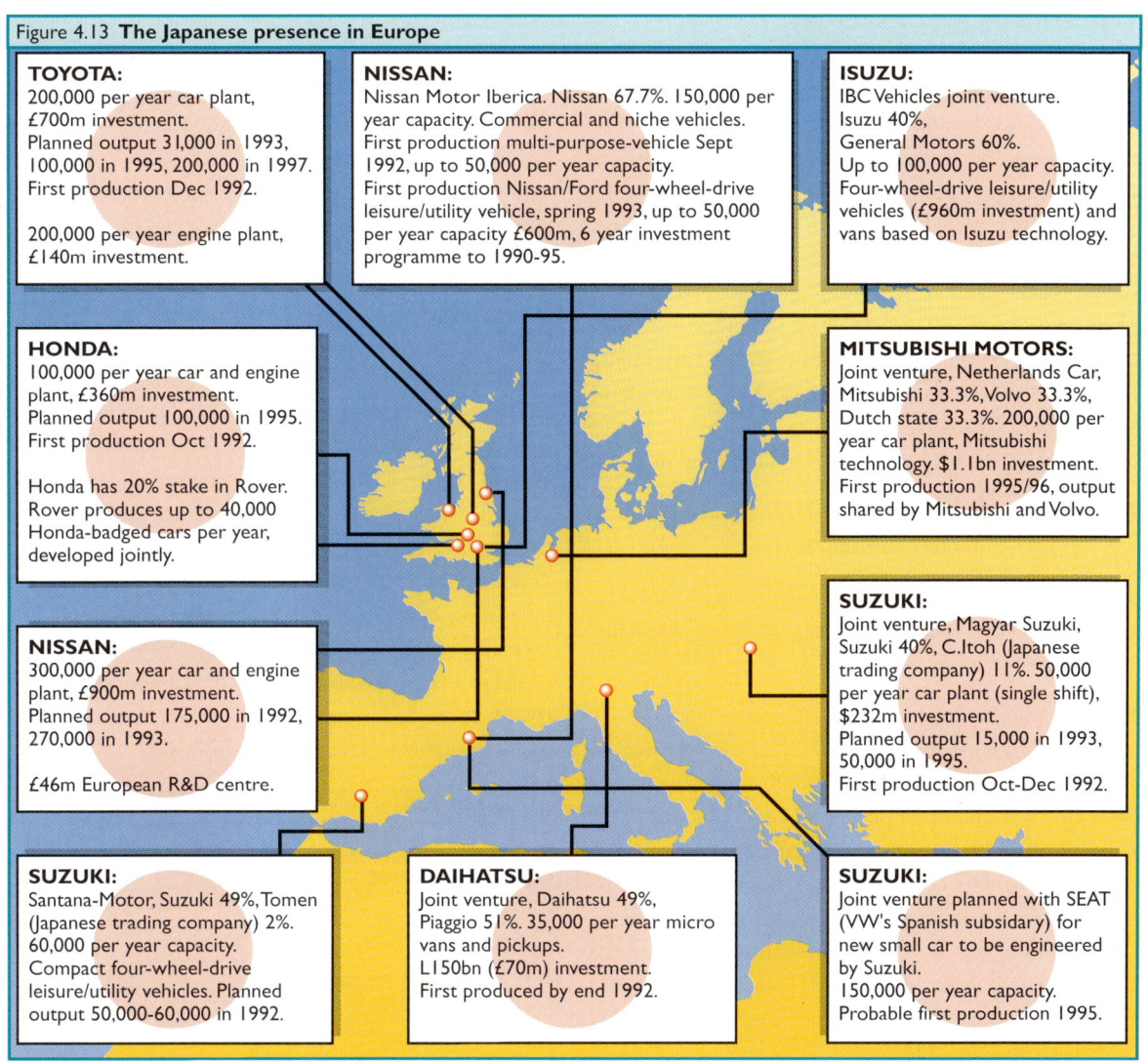

Figure 4.13 The Japanese presence in Europe

THE JAPANESE AND GLOCALISATION

Figure 4.13 shows the large Japanese presence in Europe, especially in the UK. However, the internationalisation of Japanese car producers like Toyota and Nissan has followed a different path to the globalisation of Ford. In the EU and USA, the largest Japanese manufacturers have aimed for a strategy of global localisation or **glocalisation**.

This strategy aims to set up fully integrated production, sourcing and research facilities which act independently. Unlike Ford's Mondeo, glocalisation aims to produce a specific type of car for the European market which will be different to that produced for the US or Japanese market. Globalisation involves geographic dispersion and glocalisation geographic concentration.

Japanese car makers want to produce European cars for a European market. Nissan is a case in point. Its plant in Sunderland draws 83% of its components from the EU. It makes its own engines and even has its own foundry to produce castings. The latest £30 million investment is in a facility at Sunderland to produce axles. The only major component still to come from Japan is the gearbox.

The Japanese presence is clearly a threat to European producers and there has been bitter infighting between member states. The UK, Germany and the Netherlands want to liberalise the trade barriers between Japan and the EU whereas France and Italy want Japanese cars out of Europe.

THE EUROPEAN RESPONSE

It is estimated that the **overcapacity** in the European market is 10–19%, some 2–3.5 million cars. When demand grew in the boom period of the late 1980s European producers opened new plants without closing old, inefficient ones. More recent projects include Fiat's Melfi plant, the VW plant in Zwicham, the Ford/VW plant in Setubal, Seat's plant in Martorell, Opel's factory in Eisenach and the Mercedes plant in Rastatt. The Japanese are also in place to squeeze the united European market. Their European operations have steadily grown, producing 80,000 units in 1989, 270,000 in 1992, and 450,000 in 1993, with the potential for over 1.2 million by the turn of the century. There will be two main results of this pressure:

1. **Rationalisation** of both production and workforce.
2. The rise of **mergers and acquisitions.**

The experience of the **German** car producers illustrate many of the structural and spatial trends that are influencing Europe's car makers.

Rationalisation

The most striking change in German factories has been on the workforce. From a peak of 788,000 employees in July 1991, the workforce had fallen by 150,000 by July 1994 and a further 5,000 jobs are scheduled to go. Germany has the highest labour costs in Europe (**Figure 4.14**). One response has been an increase in foreign components sourcing, using low-wage countries such as Poland and the Czech Republic. This has resulted in an increase of non-German parts built into German make vehicles from 25% in 1992 to 35% in 1994.

Another response has been to subcontract or share the burden of components research and manufacture. Mercedes Benz uses both approaches, handing over Seat production at its Bremen works to Leister Pecaro and sharing production of power steering units with its suppliers.

The urge to merge

VW has embraced the transplant factory as a means of expansion spending $51 billion on extending its operations in Mexico and exploiting the East European markets with a 70% stake in the Czech works and an effective takeover of the Spanish SEAT. BMW has also moved to establish an assembly plant in the USA. The factory at Spartanburg, South Carolina, represents a $400 million investment. They are also prepared to invest $1.3 billion in a joint venture with the Chinese government. BMW's £800 million takeover of Rover illustrates another trend in the European car industry, the **urge to merge**. By taking over Rover it has doubled its production capacity and netted the successful Rover, MG and Land Rover lines. At the same time it has gained a low cost base in Britain.

The future

The European Commission is basing its strategy for the future of the car industry on the rather ungainly slogan 'clean, lean-produced, intelligent, quality, value.' The move towards lean production and the various global strategies will lead to plant closures and redundancies. The alliance between BMW and Rover may mean that the future is based on joint ventures and mergers. There are also opportunities for growth implied in the new production techniques. **Just-in-time** production calls for a close interaction between suppliers and assembly plants. It is estimated that by the year 2000 at least 60% of the value-added to a car will be supplied by the components industry. There are clear regional opportunities for areas to specialise in supplying components (**Inset 4.4**).

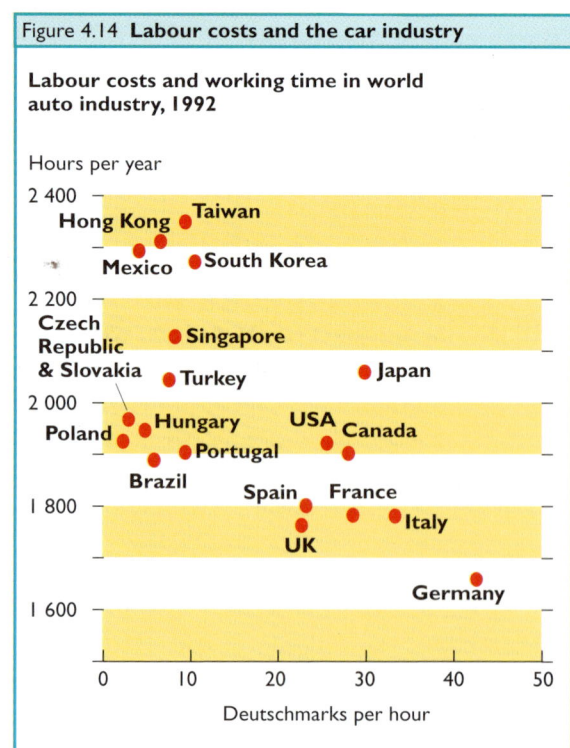

Figure 4.14 **Labour costs and the car industry**

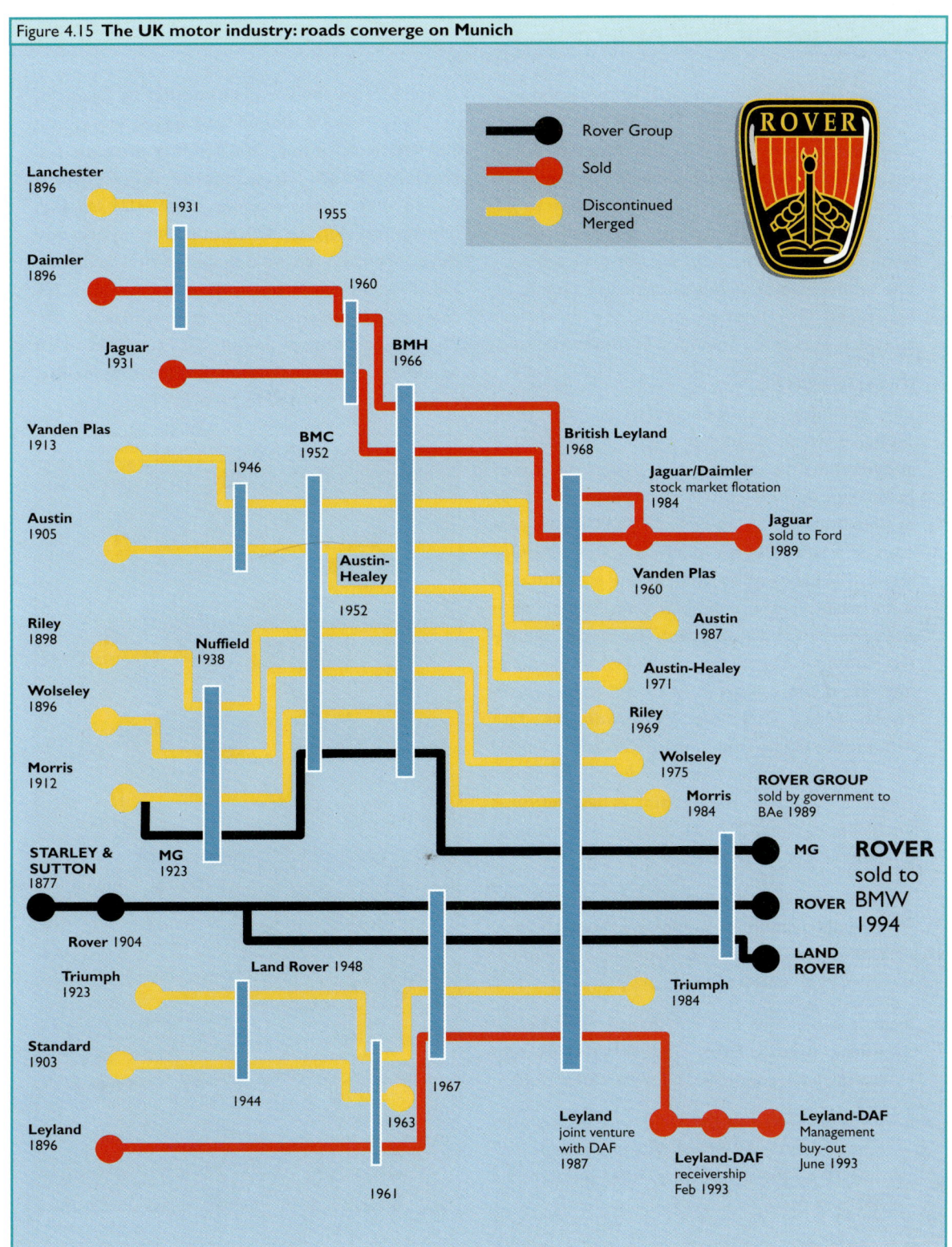

Figure 4.15 **The UK motor industry: roads converge on Munich**

INSET 4.4 JUST-IN-TIME AND LOCAL DEVELOPMENT (JIT-LED)

The concept of just-in-time (JIT) production was discussed on page 56. The JIT production discipline was introduced by the Japanese in the UK and has been eagerly adopted by European producers, especially Renault. In essence, JIT reduces the need to store parts and allows producers to be more flexible with their product. The suppliers must deliver parts just-in-time to be assembled into finished goods and producers must deliver finished goods just-in-time to be sold.

JIT-LED STRATEGY

This need for car producers to be near their suppliers which is implied by JIT has led some geographers to suggest the development of growth-pole like spatial concentrations or agglomerations. The so-called JIT-LED strategy is seen as a new tool for local economic development. JIT-LED may be centred on a traditional automobile centre, a branch plant region or a site of new inward investment. The Black Country Development Corporation in the West Midlands has seized the opportunity to sell its region in terms of all three to JIT producers. The Automotive Component Park (ACP) has adopted the slogan 'they built the motor industry around us'. Figure 4.16 shows their claim is not without justification. The ACP has a number of locational advantages for JIT producers:

1. It is one hour's drive by HGV to the car assembly plants of Jaguar, Peugeot, Reliant, Rolls-Royce, Rover and Toyota which produce more than half a million cars per year.

2. It is within two hours of plants operated by Rover, Toyota, Ford, General Motors and Honda which construct a further 500,000 cars.

3. It is well linked to the UK's road network by junctions 1 and 2 of the M5, and junctions 9 and 10 of the M6, with further links to the M1 and M40.

These factors, plus a combination of financial assistance and a skilled and eager workforce has attracted many components firms to the West Midlands. They include the German-owned vehicle lock manufacturer, HUF UK Ltd, and Versan Wilkins which supplies Toyota and Rover from its £6 million plant at Darleston. However, the ACP has only attracted one US company, Johnson Controls Automotive, so far. The US corporation invested £12 million in an 81,000 square foot foam manufacturing facility, which was completed in June 1994.

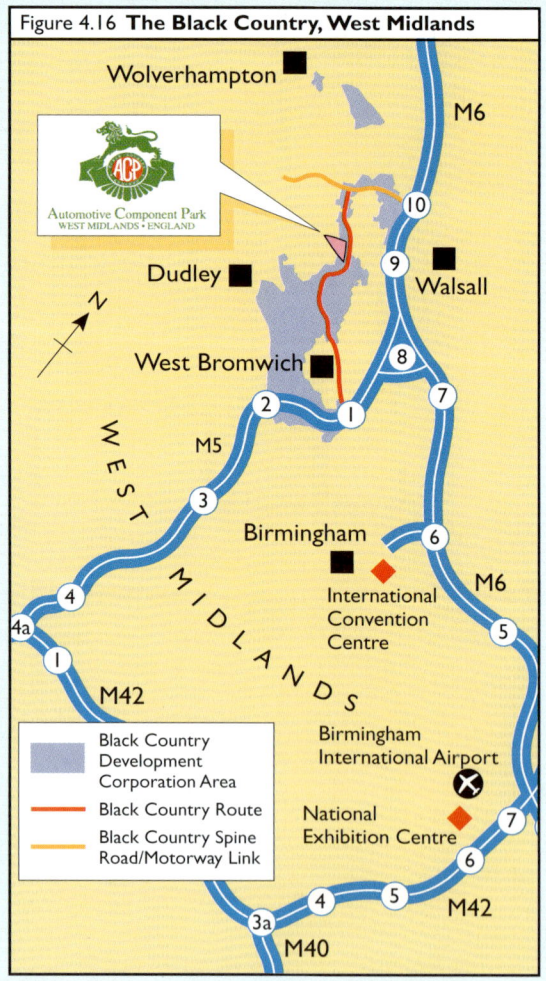

Figure 4.16 The Black Country, West Midlands

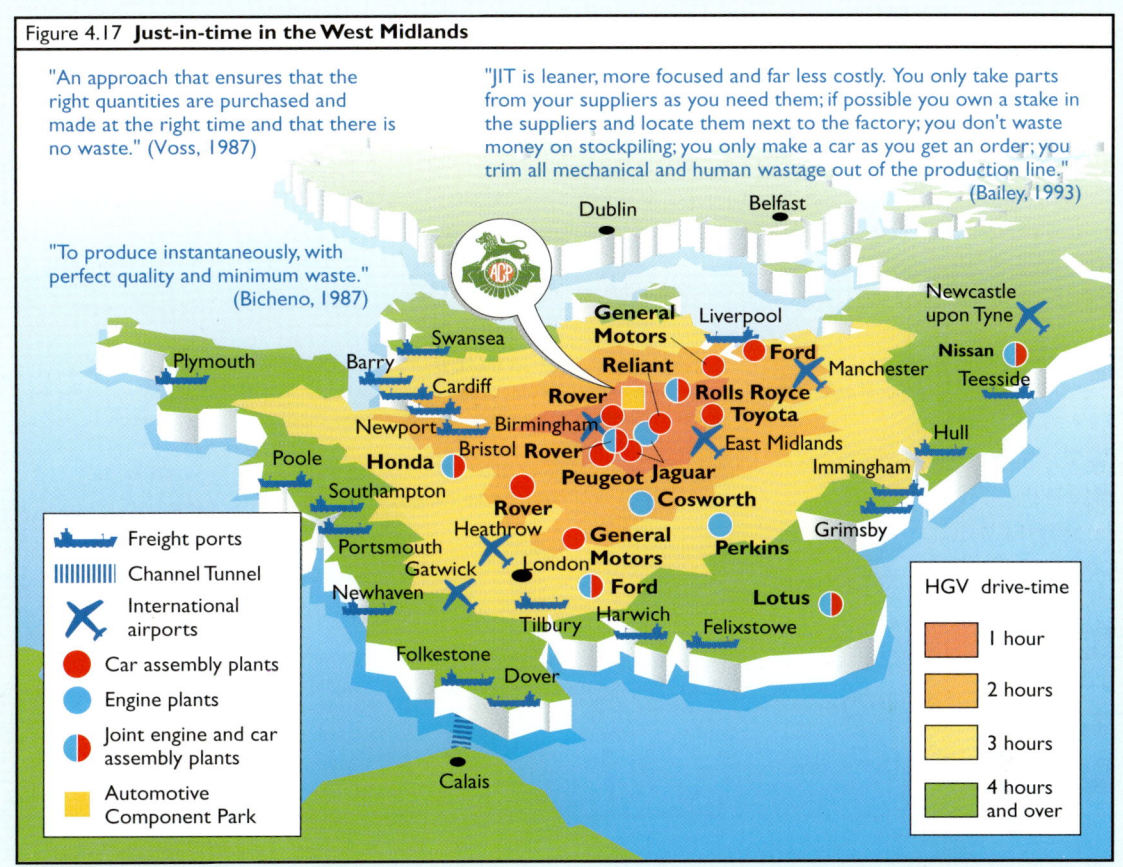

Figure 4.17 **Just-in-time in the West Midlands**

"An approach that ensures that the right quantities are purchased and made at the right time and that there is no waste." (Voss, 1987)

"JIT is leaner, more focused and far less costly. You only take parts from your suppliers as you need them; if possible you own a stake in the suppliers and locate them next to the factory; you don't waste money on stockpiling; you only make a car as you get an order; you trim all mechanical and human wastage out of the production line." (Bailey, 1993)

"To produce instantaneously, with perfect quality and minimum waste." (Bicheno, 1987)

JIT-LED ASSESSED

Recently the influence of JIT on local economic development has been questioned. It has three perceived benefits:

1 Growth pole development through employment creation.
2 Multiple local spin-offs in JIT-related activities.
3 Investment by JIT manufacturers in training and supplier company infrastructures which benefits the local community.

On the other hand, there are also problems associated with stimulating JIT investment:

1 The fact that friction of distance has been overemphasised in the JIT-LED concept: same day delivery could quite feasibly include most of the EU for the assembly plants of GM, Ford and Rover in southern England.

2 The tendency towards company regions in which JIT manufacturers exercise unwanted influence in local affairs.

3 The amount of money which is spent in regional assistance and advertising to attract industries to an area.

What is clear is that JIT, lean production and the increasing Japanese influence on the car industry looks set to influence the location of assembly and component plants in areas like the West Midlands well into the next century.

Section F Exercises and recommended reading

EXERCISES

1. Define the following terms: *globalisation, glocalisation, just-in-time production,* and *Fordism.* [8]

2. **Figure 4.18** shows the major European car producers. Use it to answer the following questions.
   (a) Describe the similarities and differences in location and markets between the European producers. [7]
   (b) What are the advantages and disadvantages of pursuing strategies of globalisation and glocalisation for the major companies. Use examples from the chapter to illustrate your answer. [15]

Figure 4.18 Globalistion or glocalisation in the car industry in the year 2000					
	MARKETS				
LOCATION	**LOCAL** national organisation	**REGIONAL I** presence in one region (exports)	**REGIONAL II** present in two or three major regions	**GLOBAL** independent dealers and importers	**GLOCAL** fully controlled local dealers
LOCAL >80% domestic	Chrysler	PSA	Rover (this may change after merger)		BMW
REGIONAL <70% local >80% regional		Renault Fiat	Volvo	VAG (VW, Audi, Seat, Skoda)	Daimler Benz
GLOBAL >70% regional			GM (including Saab)	Ford (including Jaguar, Mitsubishi)	Mazda Honda
GLOCAL major presence in USA, Japan, Europe					Toyota Nissan

3. (a) Graph the data given in **Figure 4.19**. [4]
   (b) Use your graph to describe and explain the success of the restructuring between 1989 and 1994. [6]

Figure 4.19 Rationalisation and restructuring of British Steel between 1989 and 1994					
	1989/90	**1990/91**	**1991/92**	**1992/93**	**1993/94**
Capital expenditure (£m)	450	459	255	197	104
Average number of employees (000s)	54	57	52	46	41
Man hours per tonne of liquid steel	4.8	4.8	4.4	4.3	3.8
R&D expenditure (gross in £m)	35	35	34	34	39

4. 'Other European Community producers need to take the same action that the British Steel industry took to reduce capacity. [The Commission] should not prop up inefficient companies by means of state subsidies, thus putting efficient, unsubsidised companies like BS at an unfair disadvantage.' [Tim Sainsbury, Minister for Industry, February 1993]
   Essay: What are the arguments for and against the use of subsidies in the European steel industry? [25]

RECOMMENDED READING

'Europe's car industry in the 1990s' (1994) in *Geography Review,* vol. 6, no. 5, pp. 26-30 and 'Trends in UK Manufacturing' (1996) in *Geofile* 275. Peter Dicken (1992) *Global Shifts*, Paul Chapman Publishing.

CHAPTER 5
HIGH-TECHNOLOGY AND SERVICE INDUSTRIES

A **HIGH-TECHNOLOGY AND KNOWLEDGE-BASED INDUSTRIES**
   Locational characteristics: global switching **83**
   Locational characteristics: European research and technological development **84**
   Locational characteristics: national division of labour **86**

B **SERVICE INDUSTRIES**
   Types of services **88**
   The growth of services **88**
   Service employment in Europe **90**
   Producer services **92**
   Retail developments **94**
   Tourism **96**

C EXERCISES AND RECOMMENDED READING **98**

Section A shows how the electronics and high-tech industries act as a link between manufacturing and service industries. The majority of the workforce are involved in routine production-line employment but there is also a highly skilled, well paid managerial and professional section within the workforce. The location of high-tech industries is linked with the development of science parks, improved communications, access to universities and/or government research labs, and a pleasant environment. Examples in Britain include the Warwickshire Science Park and Siemens in the North East. High-tech industries are closely linked with producer services: their growth rates and contribution to the economy are impressive, their products have diversified and markets have expanded. Although some services share these characteristics not all do.

Section B examines the service sector showing its diversity of employment from insurance, banking, shipping, tourism, health care, refuse collection, entertainment and education to retailing. This effectively covers all economic activities other than the production of goods. However, there is great diversity in the organisation of service industries: some services are provided by the state, such as the National Health Service in the UK, whereas others are provided by the private sector, such as market research and advertising. Some jobs, such as international banking, are extremely well paid, while others are very poorly paid, notably refuse collection. Many jobs are traditionally female, including catering, and cleaning, while others are traditionally male, such as transport.

In the post-industrial society that characterises Europe service industries are increasingly important for regional development. They provide employment, exports and affect a country's balance of payments. There are also indirect benefits, through employees expenditure in the local economy. In general, with the exception of tourism, major service activities tend to occur in the leading city regions, such as London, Paris, Brussels, Hamburg and the Randstad.

Section A High-technology and knowledge-based industries

High-technology industries include a large variety of firms and processes. The term 'high-technology' is often used to refer to the microelectronics industry. However, a microelectronics assembly plant may be relatively low-technology and low value added, involving little more than semi-skilled workers soldering semiconductors to circuit boards. Whereas, a car plant which uses sophisticated robotics is not classed as a high-technology firm even though it does depend on techniques developed by high-technology firms. For these reasons the terms knowledge-based and brain-intensive are increasingly in use around Europe. Such industries are characterised by:

1. High inputs of scientific research and technological development (R&D).
2. Rapid technological change with ever shortening product life cycles and a rapid growth in market demand.
3. Innovative and technologically advanced products.
4. Strong linkages between large multinational corporations and smaller 'boffin' enterprises.
5. A highly qualified workforce with a high proportion of professional engineers and scientists.

The classic high-technology industries include computers, silicon chips, lasers, robotics, aerospace and medical equipment but there is an equally strong case to include pharmaceuticals, biotechnology and telecommunications equipment.

The Triad

High-technology industries represent the leading edge of development in most industrialised countries. In terms of research and development, the EU seems well established. Within the member states the industries with levels of expenditure in excess of 10% of total turnover are all knowledge-based industries. In rank order they are electrical goods, electronics, aeronautics and pharmaceuticals. However, **Figure 5.1** shows a general weakness when compared with the USA and Japan, together with which Europe forms a **triad** of leading trading groups.

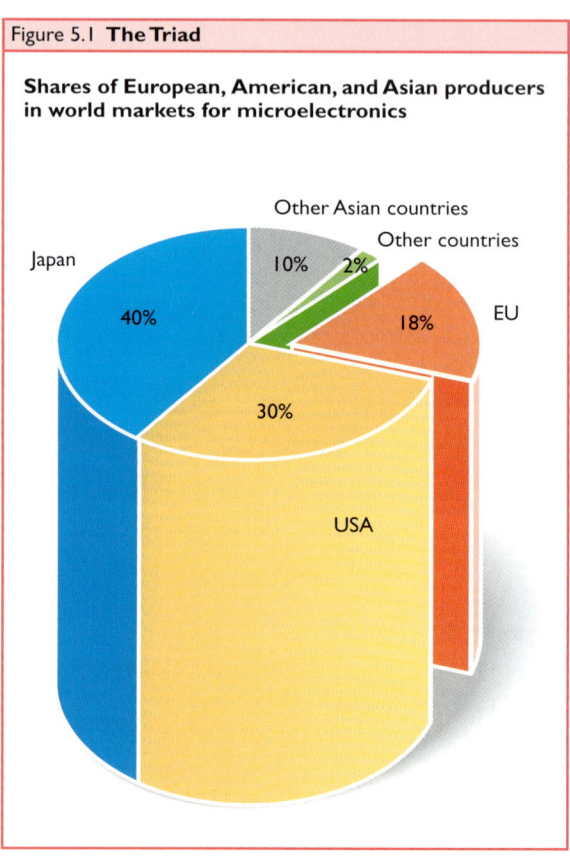

Figure 5.1 **The Triad**

Shares of European, American, and Asian producers in world markets for microelectronics

For example, the **European electronics industry** is anything but thriving. Since 1974, the relative place of the European industry in the world has declined as innovation of new products has been increasingly determined by US and Japanese producers. The European industry has also suffered the indignity of successful foreign mergers such as Fujitsu's takeover of ICL and Nokia. There is a trading deficit in electronics between Europe and the rest of the world which widened from $1.5 billion in 1979 to over $45 billion in 1993. This is particularly apparent in those areas of electronics which could be best described as high-technology: computers, consumer electronics and microchips. **Figure 5.1** shows that there is still a triad of producers—Europe, the USA and Japan—but it also shows the European industry is lagging behind.

This section now looks at the locational characteristics of European high-technology firms at the **global**, **European** and **national** levels.

LOCATIONAL CHARACTERISTICS: GLOBAL SWITCHING

The idea of the **globalisation** of industries is a common theme in this book, first discussed in **Chapter 1**. Globalisation entails the international expansion and integration of key corporate functions such as **production, marketing** and **research and technological development**. It also implies the growth in international collaboration and linkages with other firms. In high-technology industries the concept of **global switching** has been suggested to explain the international locations of large firms.

The essence of global switching is the ability of companies to coordinate their different function operations (i.e. R&D, manufacture, marketing, sales administration) in an integrated fashion on a global scale. This can be seen in a **spatial division of labour** at a global scale and in a movement of research and development facilities away from a multinational corporation's mother country to a new market abroad. For example, Japanese firms are establishing overseas R&D centres in Europe: in 1990, NEC put in operation a semiconductor research centre at Dusseldorf (Germany), Hitachi a research centre on basic technology for submicron production at Cambridge (UK), and Fujitsu a design centre for VLSI circuits in London.

Case Study: Glaxo

The influence on industrial location is best seen with the development cycle of a new product from initial discovery and invention through to the first market launch. **Figure 5.2** provides a broad outline of a new anti-asthma drug developed by Glaxo—clearly a knowledge-based, brain-intensive process. A sevenfold global switch was the result:

1. The product, *Salmetrol*, involved R&D at its site at Ware in the UK.
2. This was followed by extensive clinical trial studies across Europe and worldwide.
3-4. The scale-up of production for the active ingredients of the drug was transferred to Montrose in Scotland, which then moved into full-scale production.
5. Another primary production site was set up in Singapore.
6. The drug was then sent to secondary production and packaging operators in Evreux, in France, and in Ware.
7. The first market launch for the product in 1990 was in the UK although other European countries are now also product launch sites.

Links to the product life cycle model

The idea of global switching has clear links to the **product life cycle model** which is discussed in **Chapter 3**. The model suggests that manufacturing and research capacity gradually shift from innovative domestic markets to foreign markets as products move from an **early** or **development** phase through to a **growth** stage and on to a **mature** phase of their life cycles. In the model, location is determined by the quest for new markets and/or cheap labour. However, global switching involves a more federated structure. As the example of **Glaxo** shows the switching to a European location occurred in the development stage and happens because:

- flows of information about **marketing and research** trials can be distributed through information technology from the parent company;
- a new product may be targeted for the particular market using **local knowledge**;
- a European location may avoid **tariff barriers**;
- European managers have **bottom-up** input with respect to marketing and other product requirements which could not be obtained in a location outside Europe.

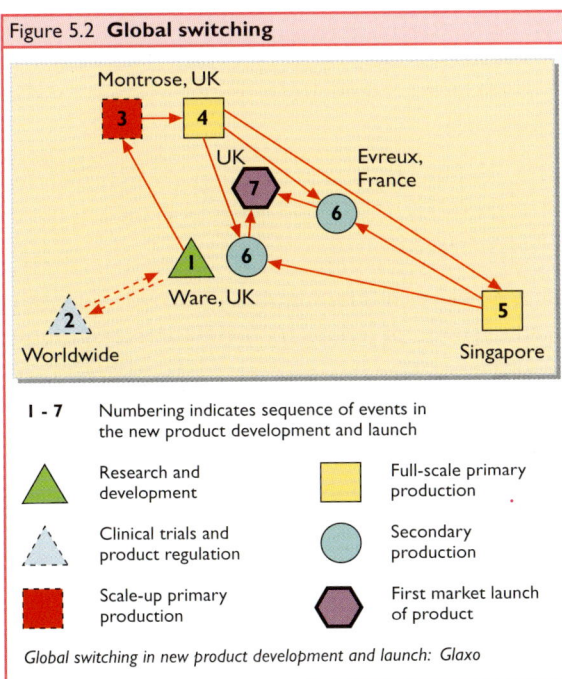

Global switching in new product development and launch: Glaxo

LOCATIONAL CHARACTERISTICS: EUROPEAN RESEARCH AND TECHNOLOGICAL DEVELOPMENT

At the European scale a number of location characteristics are evident. The influence of global switching has already been noted but there are other factors which affect the location of foreign and European firms, both large and small.

Core and periphery

One of the best indicators of high-technology activity is expenditure and employment which a region devotes to **research and technological development (R&D)**. Every region in Europe wants to develop R&D activities as the capacity to innovate and upgrade processes allows for flexibility of production, high value added products and a better trained workforce. Moreover factors which favour R&D are virtually the defining characteristics of the EU's core regions:

- well developed communications networks;
- good scientific infrastructure;
- access to a skilled and educated workforce;
- advanced markets for business and information services.

Figure 5.3 shows the concentration of R&D labour force in Europe's core regions. The problems for weaker regions are twofold: to generate and develop their own indigenous R&D activities and to adapt technological developments which take place elsewhere to a specific regional context. The result is that many peripheral regions attempt to attract foreign investors. Research has shown that, on average, foreign-owned companies in Spain and Ireland are more likely to conduct research activities than native firms. **Figure 5.3** shows that with respect to R&D labour force Greece and Portugal have less than a fifth of the scientists and engineers per 1000 employed than the more advanced member states. However, there are exceptions. **Inset 5.1** on **page 86** describes Siemens' decision to locate in Tyneside. The plant includes a research division.

Islands of innovation

The most striking locational characteristic of high-tech firms which are involved in R&D projects is their concentration in comparatively few '**islands of innovation**' (**Figure 5.3**). The islands are relatively small, well linked areas, with a dense network of enterprises and research laboratories. Once again, the links to the core are clear. The figure shows clustering around a small number of islands: Greater London, Rotterdam/Amsterdam, Ile de France, the Ruhr area, Frankfurt, Stuttgart, Munich, Lyon/Grenoble, Turin and Milan. Up to three-quarters of all public research contracts, including those funded by the EU, are concentrated in these areas. They tend to work together, with linkages ensuring a rapid exchange of ideas through collaboration.

Islands of innovation also occur in the peripheral member states and are often concentrated in a few regions, normally around capital cities. In Portugal, nearly 90% of public sector R&D is carried out in Lisbon and the Tagus Valley. In Italy only 3% of industrial research undertaken by the private sector takes place in the South and barely 9% of public sector research.

European science parks

A key influence on the location and success of Europe's islands of innovation are science parks. These parks, whether they be in the UK or continental Europe, frequently involve partnerships between a university, a local authority, a development agency and sometimes company partners. They are defined as a commercial property enterprise which has:

1. Formal and operational links with a university or other higher educational institution or major centre of research.

2. A structure designed to encourage the formation and growth of knowledge-based industries and other high-tech organisations.

3. A management function which is actively engaged in the transfer of technology and business skills to the organisation on site.

The parks are generally located in areas of traditional industry which, in varying degrees, have suffered from recession. A common objective in establishing these science parks is to provide a catalyst to help to change the industrial structure of a region (see the Ruhr, **pp. 128-31**).

They tend to be located on **greenfield** sites in pleasant, out-of-town locations although they have also been used to stimulate inner city areas, e.g. the successful Manchester

Figure 5.3 Research and development personnel and islands of innovation

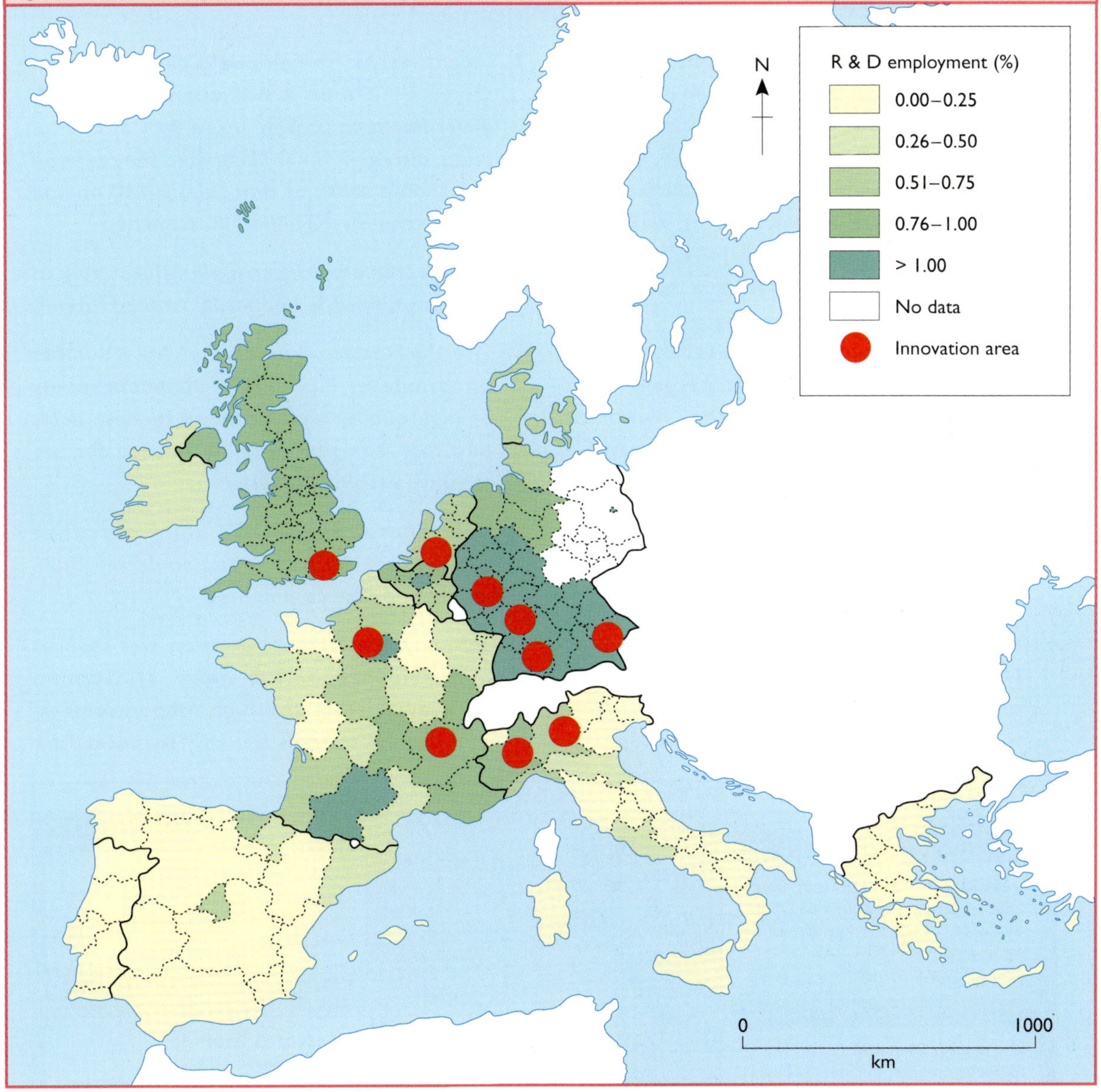

Science Park situated in the inner city borough of Hulme. Nationally, the parks have a dual function of marketing a university's expertise and providing an incubator for small high-technology firms. Across Europe they develop **linkages** which enable exchange of ideas throughout the EU. Inset 5.1 provides an example of the way science parks can achieve this.

The benefits of science parks are clear and exist at a number of levels:

1. **Business**, including the transfer of technology, use of university equipment, agglomeration economies.
2. **University.** Provides revenue for university, allows students and academics real world experience, marketing of university research.
3. **Region.** Local economic development through direct and indirect employment, the strengthening of local industry, diversification of industry, attraction of inward investment.

INSET 5.1 LOCATIONAL CHARACTERISTICS: NATIONAL DIVISION OF LABOUR

SIEMENS: LOCATING IN TYNESIDE

In August 1995, Siemens, the German electronics group, announced that it had chosen the Hadrian Business Park, North Tyneside, to build its £1.13 billion semiconductor manufacturing and research centre. The investment, which will create approximately 2,000 jobs, was won against fierce competition from a number of other European locations (Figure 5.4). The decision was a coup for the recently formed English Partnerships, the economic regeneration agency for England, and the Northern Development Corporation (NDC), who provided Siemens with an attractive package of cash incentives, labour and site characteristics.

These include:

1. **Research opportunities**: the park has four universities within 30 minutes drive, notably Newcastle, which has a world-class computing and electrical engineering department, specialising in semiconductor technology.

2. **Incentives**: North Tyneside is classified as an Objective 2 area which allows the British government to provide up to 30% of capital costs, although the DTI claimed they offered less than a tenth of that (about £50 million of the total £1.13 billion investment).

3. **Labour**: the area offers a plentiful supply of educated, flexible and easily trained labour.

4. **Site**: the Hadrian Business Park has a number of advantages, including an abundant supply of high quality water needed to rinse acids and the geological stability needed for the sensitive etching process.

5. **Behavioural**: the NDC's chief executive spoke fluent German—'You could see them visibly relax' he commented.

In Germany, the Siemens decision was seen as further evidence of the move of German industry away from the high labour costs of their homeland towards the UK (see Inset 1.5).

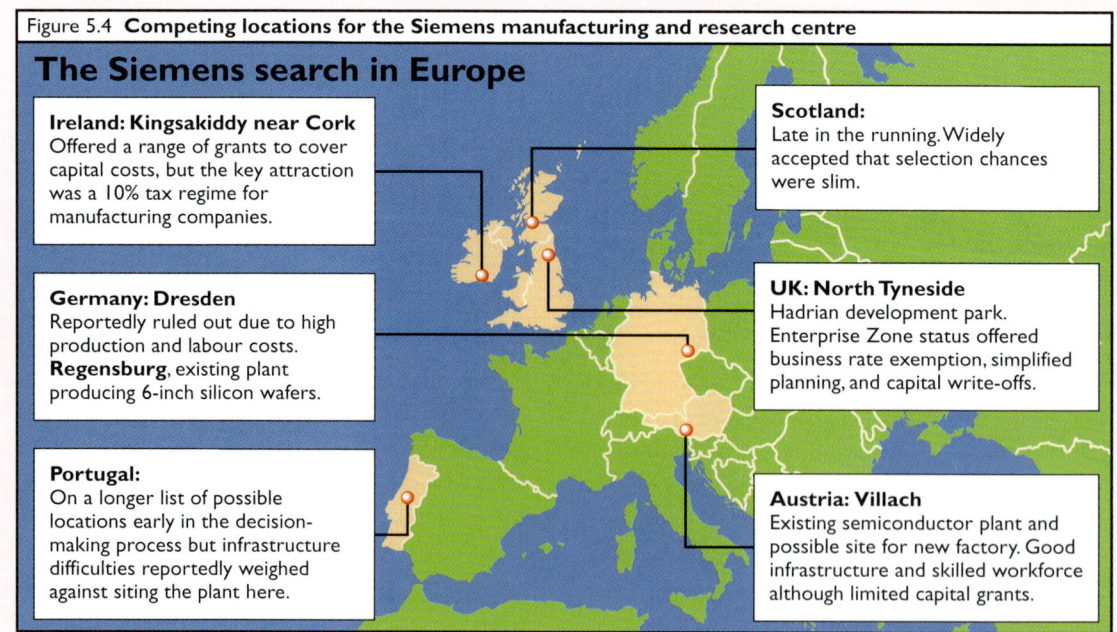

Figure 5.4 Competing locations for the Siemens manufacturing and research centre

The Siemens search in Europe

Ireland: Kingsakiddy near Cork
Offered a range of grants to cover capital costs, but the key attraction was a 10% tax regime for manufacturing companies.

Germany: Dresden
Reportedly ruled out due to high production and labour costs.
Regensburg, existing plant producing 6-inch silicon wafers.

Portugal:
On a longer list of possible locations early in the decision-making process but infrastructure difficulties reportedly weighed against siting the plant here.

Scotland:
Late in the running. Widely accepted that selection chances were slim.

UK: North Tyneside
Hadrian development park. Enterprise Zone status offered business rate exemption, simplified planning, and capital write-offs.

Austria: Villach
Existing semiconductor plant and possible site for new factory. Good infrastructure and skilled workforce although limited capital grants.

HIGH-TECHNOLOGY AND KNOWLEDGE-BASED INDUSTRIES

Figure 5.5 SPRINT partnerships

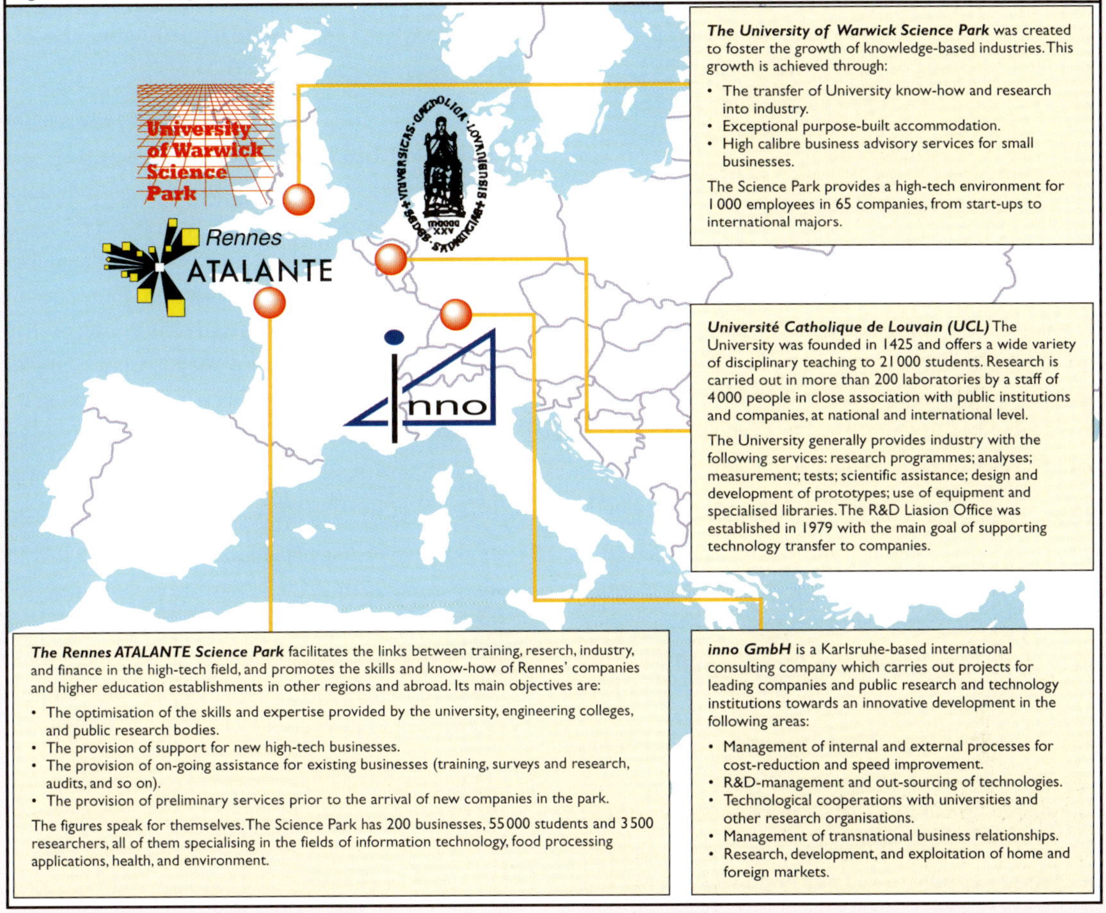

The University of Warwick Science Park was created to foster the growth of knowledge-based industries. This growth is achieved through:

- The transfer of University know-how and research into industry.
- Exceptional purpose-built accommodation.
- High calibre business advisory services for small businesses.

The Science Park provides a high-tech environment for 1 000 employees in 65 companies, from start-ups to international majors.

Université Catholique de Louvain (UCL) The University was founded in 1425 and offers a wide variety of disciplinary teaching to 21 000 students. Research is carried out in more than 200 laboratories by a staff of 4000 people in close association with public institutions and companies, at national and international level.

The University generally provides industry with the following services: research programmes; analyses; measurement; tests; scientific assistance; design and development of prototypes; use of equipment and specialised libraries. The R&D Liasion Office was established in 1979 with the main goal of supporting technology transfer to companies.

The Rennes ATALANTE Science Park facilitates the links between training, reserch, industry, and finance in the high-tech field, and promotes the skills and know-how of Rennes' companies and higher education establishments in other regions and abroad. Its main objectives are:

- The optimisation of the skills and expertise provided by the university, engineering colleges, and public research bodies.
- The provision of support for new high-tech businesses.
- The provision of on-going assistance for existing businesses (training, surveys and research, audits, and so on).
- The provision of preliminary services prior to the arrival of new companies in the park.

The figures speak for themselves. The Science Park has 200 businesses, 55 000 students and 3 500 researchers, all of them specialising in the fields of information technology, food processing applications, health, and environment.

inno GmbH is a Karlsruhe-based international consulting company which carries out projects for leading companies and public research and technology institutions towards an innovative development in the following areas:

- Management of internal and external processes for cost-reduction and speed improvement.
- R&D-management and out-sourcing of technologies.
- Technological cooperations with universities and other research organisations.
- Management of transnational business relationships.
- Research, development, and exploitation of home and foreign markets.

WARWICK SCIENCE PARK

There are 200 science parks in the EU. They are now linked together by the fast electronic data communications of the internet. The EU is keen to encourage linkages between science parks in different member countries. For example, the UK Science Park Association has three members outside Britain, from the Netherlands, Norway and Belgium.

The University of Warwick Science Park was established by the university together with Coventry City Council and Warwickshire County Council in 1984. Companies range from international majors such as Westinghouse, Computervision, Olivetti and Coopers & Lybrand to many new start-ups. The park is part of the SPRINT (Strategic Programme for Innovation and Technology Transfer) network, which is an EU subsidised programme designed to help small and medium sized companies to develop partnerships either for business or technology transfer. The network is led from Rennes in Brittany, with partners in Louvain in Belgium and Karlsruhe in Germany. Links include biomedicine and robotics, with more than 70 firms linked in the network overall.

Section B Service industries

TYPES OF SERVICES

1   **Producer** services are 'high order' activities such as market research, management consultancy, financial, advertising and legal functions, that are provided in a small number of highly developed metropolitan centres, generally capital cities, for other firms or organisations.
    Consumer services are provided generally for people, e.g. health care, retailing, education, distribution and refuse collection. These are more local in scale or 'low order'.

2   **Non-basic** services are provided for users in the local area and generally have little multiplier effects.
    Basic services are those which are provided for a market beyond the local economy, e.g. a national or a global market, hence bringing money into a region.

3   **Private** or **market** services are organised by independent companies, ranging from contract-cleaners and retailing to international banking and insurance.
    State or **non-market** services are organised by the government such as central and local government, the National Health Service and education.

In much of the literature a threefold division is commonly adopted: high order producer services, low order consumer services (**Figure 5.6**) and tourism. Variations in these types of services are important in determining whether a region will develop successfully in the post-industrial society. For example, an area which depends upon **consumer non-basic services**, such as post offices, schools or supermarkets, will attract little extra growth (multiplier effect), whereas **producer basic services**, such as international financial institutions, generate significant multiplier effects and contribute to the region's and the nation's economic growth. Moreover, there can be a transition from consumer services to producer services. A small bank or building society might only serve a local (non-basic) market. If, as in the case of the Halifax Building Society, it expands and/or merges (in this case with the Leeds) to serve a national (basic) market, it has effectively evolved from a consumer service to a producer service. Similarly, the growth of retailing and travel firms, such as Sainsbury's and Thomas Cook, illustrates the change from local to international organisations.

THE GROWTH OF SERVICES

The growth of the service sector, or **tertiarisation**, was described in **Clark's sector model** (p.10). A number of reasons have been suggested:

1   It has traditionally been explained by **increasing efficiency** and **productivity** of **manufacturing** sectors thereby giving employees a larger disposable income. This would be spent on services such as leisure, transport or education. The corresponding decline of manufacturing employment, **deindustrialisation**, due to increased mechanisation, increased foreign competition or changes in demand, has forced many manufacturing employees into the service sector. Another partial explanation is that consumers in developed countries have been buying less home-produced goods and more overseas goods, due to increased competitiveness.

2   As manufacturing firms have **rationalised** they have sub-contracted much of their internal services to the private sector, such as catering, accountancy, cleaning and market research. These service jobs, traditionally classified as 'manufacturing' since they were carried out internally, are now classified as 'service employment'.

3   **Demographic** changes may lead to increased demands for service functions: an ageing population, for example, will require greater amounts of health care.

4   **New technology** also increases the demand for services: the growth of cash dispensers and cash cards has only occurred since the 1980s. Moreover, it is sometimes argued that services are not capable of the same improvements in productivity as manufacturing is, thus employment levels remain high. However, the evidence does not necessarily back this view and it is true that the growth of the service sector has varied considerably in terms of type of service and time of growth.

None of these explanations is adequate on its own. It is necessary to evaluate them in relation to particular places and to consider such forces as the globalisation of manufacturing and service industries, trade and organisation, deregulation of markets, growing specialisation of service outputs, increasing flexibility of production and advances in communications.

Figure 5.6 **Service employment: London's futures exchange, day and night**

SERVICE EMPLOYMENT IN EUROPE

The map of service employment in Europe is rather complex. A number of regions show extremely high rates of service employment: the South East of the UK, Ile de France, Brussels, Hamburg, west Netherlands, Madrid, Lazio (Central Italy), the French Mediterranean and the Canaries. Moreover, there are high rates in such diverse areas as Northern Ireland, Scotland, the South West of the UK, Denmark, Walloon (Belgium), Attiki (Greece) and parts of Northern Germany and the Netherlands. The explanation lies in the type of service provided. The regions can be divided into three broad categories:

1 **Producer services** centres such as London, Paris, Hamburg and Brussels.
2 **Tourist and/or retirement** locations such as the South West, the Canaries, French Mediterranean and Greece.
3 Nation-regions such as Scotland, Northern Ireland and Denmark (but not Ireland).

The numbers employed in services in Europe increased by over 50% between 1971 and 1990. The main reason for variations in the growth rate reflects the type of services present in each country. **Fastest growth** rates were in the **producer services**, notably finance and business and also the Mediterranean regions, which experienced rapid growth in services related to tourism. By contrast, the **lowest rates of service employment** in the European Union are found in parts of the former **East Germany**, **Portugal** and **Greece**. This suggests low levels of development, i.e. concentration on manufacturing in the German regions, and on agriculture in Portugal and Greece.

Service employment in the UK illustrates many of these features. The service industry employs about seventeen million people in the UK and accounts for two-thirds of GDP. Employment is concentrated in the South East (73.8%), especially London (74.1%), and the adjacent South West (67.9%) and East Anglia (65.3%). To a lesser extent Northern Ireland and Scotland have high proportions employed in services, 66.2% and 65.4% respectively. By contrast the lowest rates of employment in service industries are found in the East and West Midlands, 59.2% and 60.1% respectively. Much of this variation can be explained by the type of service employment: London's role as a 'world city'; producer services in the South East as well as consumer services; decentralisation of functions

Figure 5.7 **A dealing room**

from the core to the surrounding areas of the South East, South West and East Anglia; tourism and an elderly population in the South West; long-term conflict in Northern Ireland, resulting in a large deployment of security forces there; and the need for special 'local' administrative services in Scotland and Northern Ireland.

In general a **north-south divide** can be identified and this has intensified during the period between 1981 and 1991. Decentralisation of producer services from London has largely been to surrounding areas. Growth in the non-basic services, such as retailing, health and education is related to population growth. This reinforces growth in the already expanding areas. By contrast, in the peripheral regions, with the exception of Northern Ireland, a decline in population and disposable income has led to a reduction in some consumer services, notably transport and retail.

However, the growth of service employment may not necessarily regenerate regions which have suffered from the losses of manufacturing industry. Although the service sector is sometimes perceived as replacing these job losses, the type of replacement service employment is vital. Consumer non-basic services will not provide much stimulus for regional development and many of the new jobs are female, low-paid and part-time.

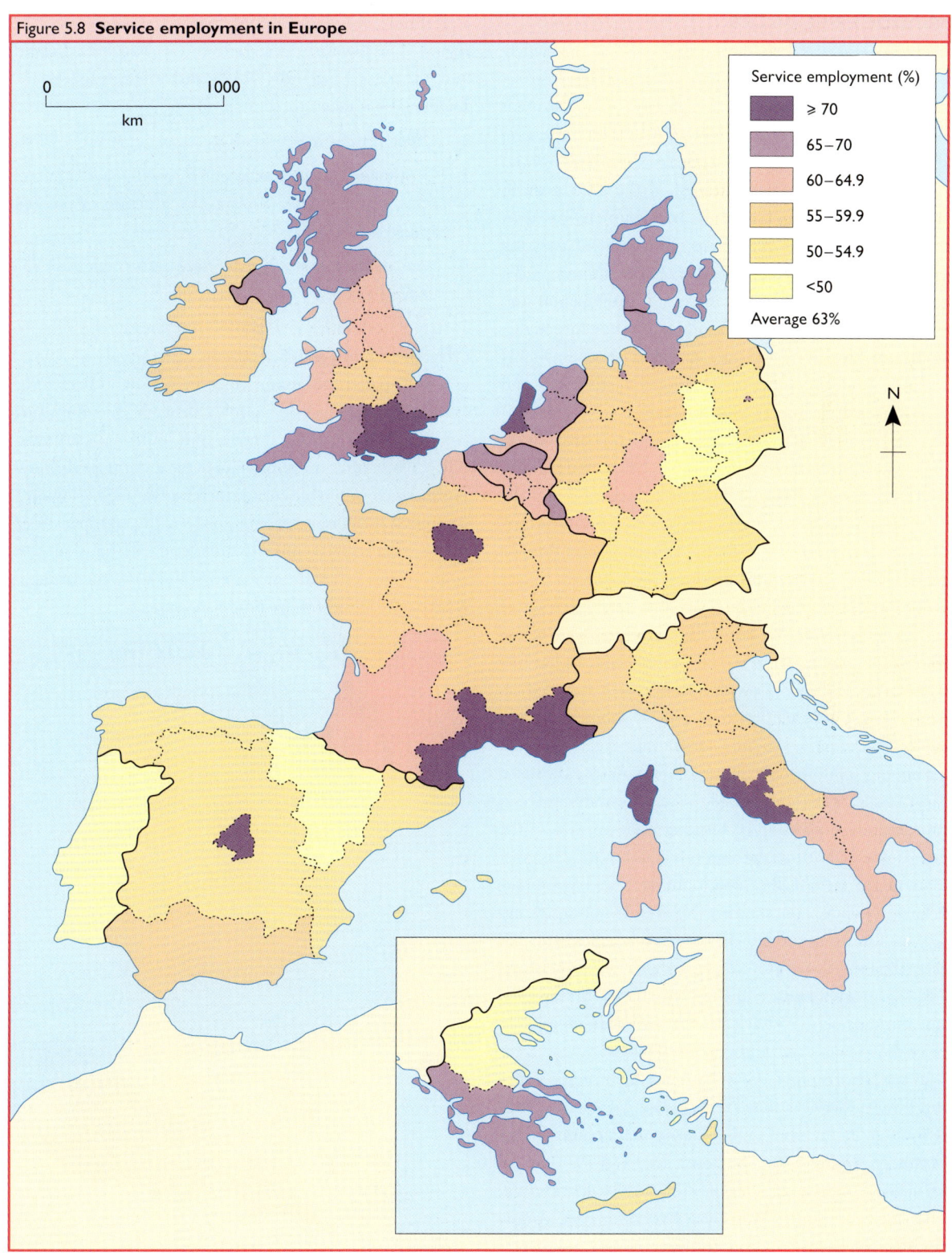

Figure 5.8 **Service employment in Europe**

PRODUCER SERVICES

Producer services are at the leading edge of the widely recognised shift from manufacturing to service employment in advanced economies. A high ratio of employment in producer services to consumer services is an indication of a favourable long-term service industry structure. Producer services are concentrated in the EU's **core** in **'world cities'** such as London, Paris, Munich/Frankfurt and Milan. For example, these cities control banking throughout Europe. **Access** to markets, customers, and a highly skilled trained labour force is a fundamental prerequisite. **Agglomeration** economies are therefore highly significant. Proximity to other specialist services, highly qualified labour, competitors, national institutions and government departments allow frequent face-to-face contacts and ensure that individual firms remain competitive.

Generally the more dynamic growth orientated regions, such as the South East, Ile de France, Randstad and the Rhein-Main, have the largest bias towards producer services. In these places there are the greatest concentrations of offices, the symbol of the producer service sector. Many new office developments intensify this concentration of high level activity. In Paris, the development of **La Défense** (**Figure 5.10**) in the north west suburbs has further reinforced disparities within Paris as well as in France. La Défense is a development of offices, shops and other services, covering 1.5 million square metres and employing about 70,000 people about 10km from the centre of Paris. Its success has intensified the east-west imbalance in Paris: deindustrialisation of working class districts in the east compared with tertiarisation in the middle-class districts of the west. Other developments include **Canary Wharf** in London and the **City-Nord** in Hamburg. A significant feature of the growth of service employment has been the even faster growth of producer services. In the UK, Germany, France and the Netherlands professional and business services have been growing at more than twice the average rate for services as a whole.

In the UK, employment in banking, finance, insurance and business services shows a very clear cut pattern: concentration in the core and under-representation in the periphery (**Figure 5.9**). The South East (18.5%), and in particular London (23.4%), dominates the distribution. The adjacent South West (12.2%) and East Anglia (10.4%) also have high proportions employed and there is a declining trend of employment away from the core region. The lowest values are found in Northern Ireland and the East Midlands, 7.1% and 7.6% respectively. **Dispersal** has taken three main forms:

1. Office clusters in suburban locations, such as Croydon.
2. Centres in locations adjacent to motorways intersections, such as Aztec West at the junction of the M4 and M5.
3. Low density landscaped office parks, such as the Oxford Science Park.

However, any move is a gamble. In general, the suburbs offer lower rents, reduced wages, easier access to labour, especially part-time labour which increasingly characterises some sectors of service employment, easier commuting conditions, lower staff turnover and improved staff reliability. These offer the **suburban areas** and the **peripheral regions** a considerable **comparative advantage**. However, prestige, a greater prospect of good quality and specialised inputs, face-to-face contacts, and good external transport connections, still make city centres and core regions highly attractive locations for producer services.

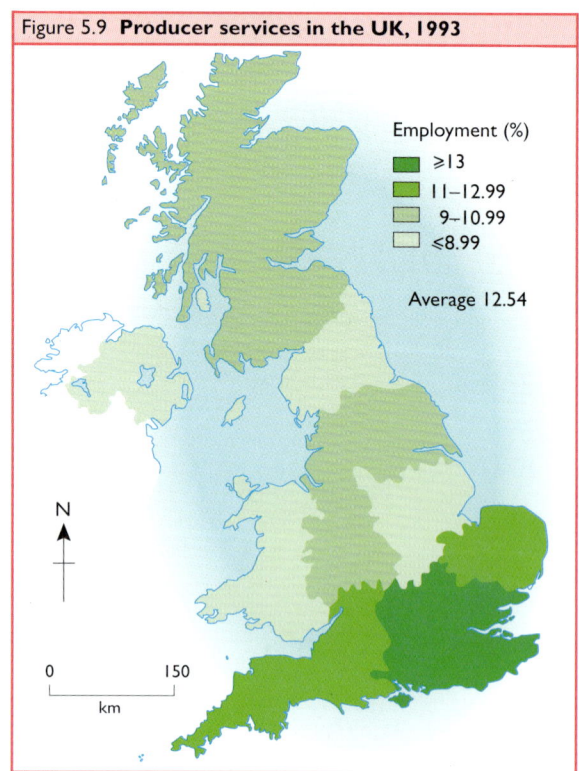

Figure 5.9 **Producer services in the UK, 1993**

Employment (%)
- ≥13
- 11–12.99
- 9–10.99
- ≤8.99

Average 12.54

Figure 5.10 **La Défense, Paris**

Producer services in Europe after 1992

The main influences in the structure of producer services after 1992 have been:

1. An integrated market.
2. Recognition of professional qualifications across borders.
3. Mergers and takeovers in the single market.

In the case of producer services the market will give firms and banks freedom to establish branches and sell financial services across national borders. Taxes and regulations that discriminate against residents of member countries will be eliminated, controls on the transfer of capital will be removed, and a common regulatory framework for financial firms will be adopted. As a result, competition will be greatly increased and prices for financial services will fall, especially in countries like Spain, where they have operated in very protected national markets.

The deregulation of banking in the EU, which came into force in 1993, allows banks from one country to operate in another. This is likely to lead to an **internationalisation** of the industry. On the other hand, banks from outside the EU have not been given the same flexibility. The same is also true of the insurance industry. The Third Insurance Coordination Directive again allows for firms of any EU member country to operate in any other member country. For example, there are seventeen UK corporate legal firms in Paris and eight in Brussels.

RETAIL DEVELOPMENTS

The retailing industry has changed from one dominated by small family firms to one in which large corporate organisations predominate. Traditionally, geographic accounts of retailing concentrated on the **location** and **type** of retailing outlet with a **central place-type hierarchy** evident. Low order goods concentrated in neighbourhood stores and shopping parades and high order goods in high street shops, department stores and, increasingly, out of town superstores and retail parks. However, modern retailing is changing rapidly.

Retailing is a major component of European economies accounting for about 13-14% of GDP in most countries. Although the number of shops has fallen from 3.5 million in 1955 to under 2.5 million in the mid-1990s the amount of floorspace has risen, hence there are fewer but larger shops. Relatively standardised retail techniques have become common in a wide variety of products and large-scale, specialist mass-merchandise outlets are widespread.

The retailing revolution has focused upon **superstores**, **hypermarkets** and **out-of-town shopping precincts** on greenfield suburban sites with good **accessibility** and plenty of **space** for parking and future expansion. The increasing use of out-of-town shopping centres, on a less frequent basis, has led to the closure of many smaller shops, which depended on frequent convenience trade. However, recent trends suggest a slow down in out-of-town developments and an increase in city-centre regeneration.

A number of factors explain the changes in retail developments:

1. **Demographic** change, e.g. falling population growth, smaller households, more elderly.
2. **Suburbanisation** and **counter-urbanisation** of more affluent households.
3. **Technological change** as more families own deep freezers.
4. **Economic change** increased standards of living, especially car ownership.
5. **Congestion and inflated land prices** in city centres.
6. **Changing accessibility** of suburban sites, especially those close to ring-road intersections.
7. **Social changes** such as more working women.

Retailers have responded in a number of ways. The most noticeable effect is a **polarisation** of retailing: many large retailers have moved into suburban and small town locations in order to capture the younger, more affluent households, whereas shops in central areas have closed. For example, the **Merry Hill Centre** near Dudley (**Figure 5.11**) has accelerated the decline of retailing in the town centre. Ten major retailers have left high street locations in order to relocate at the Merry Hill complex. On the other hand, Merry Hill is one of the most successful retailing centres in the UK. It is the UK's second largest retailing centre, containing over 167,000 square metres of retailing space and it offers retailing, banking, leisure and community services. It is the largest tourist attraction in the West Midlands, with 23.5 million visitors annually, 18.8 million of whom shop there. The Merry Hill centre contains a mixture of high street stores and specialist shops. It is built on the site of the former Round Oak steelworks, which closed in 1980. Access by car and coach is crucial to its success and it is located on the A4036 with good access to junctions 2 and 3 of the M5.

However, the perceived threat of out-of-town retailing to the town centre has resulted in a variety of attempts to maintain or revitalise retailing in central areas. This is of increasing importance and in some places this has coincided with inner city **redevelopment**. Renovation may entail full-scale development as in the case of Princes Square in Glasgow and Newcastle, or it may involve the creation of traffic free zones such as Croydon, Bruges and Cologne, or large-scale redevelopment schemes such as Les Halles in Paris or Part Dieu in Lyons. However, as a rule, redevelopment has only affected small sections of the urban core. Other developments include the covered shopping mall and specialist shopping centres, such as the Covent Garden centre and the nearby Burlington Arcade.

Different strategies for competitive advantage exist, each of which has a different spatial implication. Location and late opening hours are particularly important to local convenience stores. Close proximity or **accessibility** for consumers allows these retailers to offer the convenience which is their **comparative advantage**. Similarly, out-of-town or suburban locations for superstores combine low land prices with availability of space to provide large sales areas with extensive car parks.

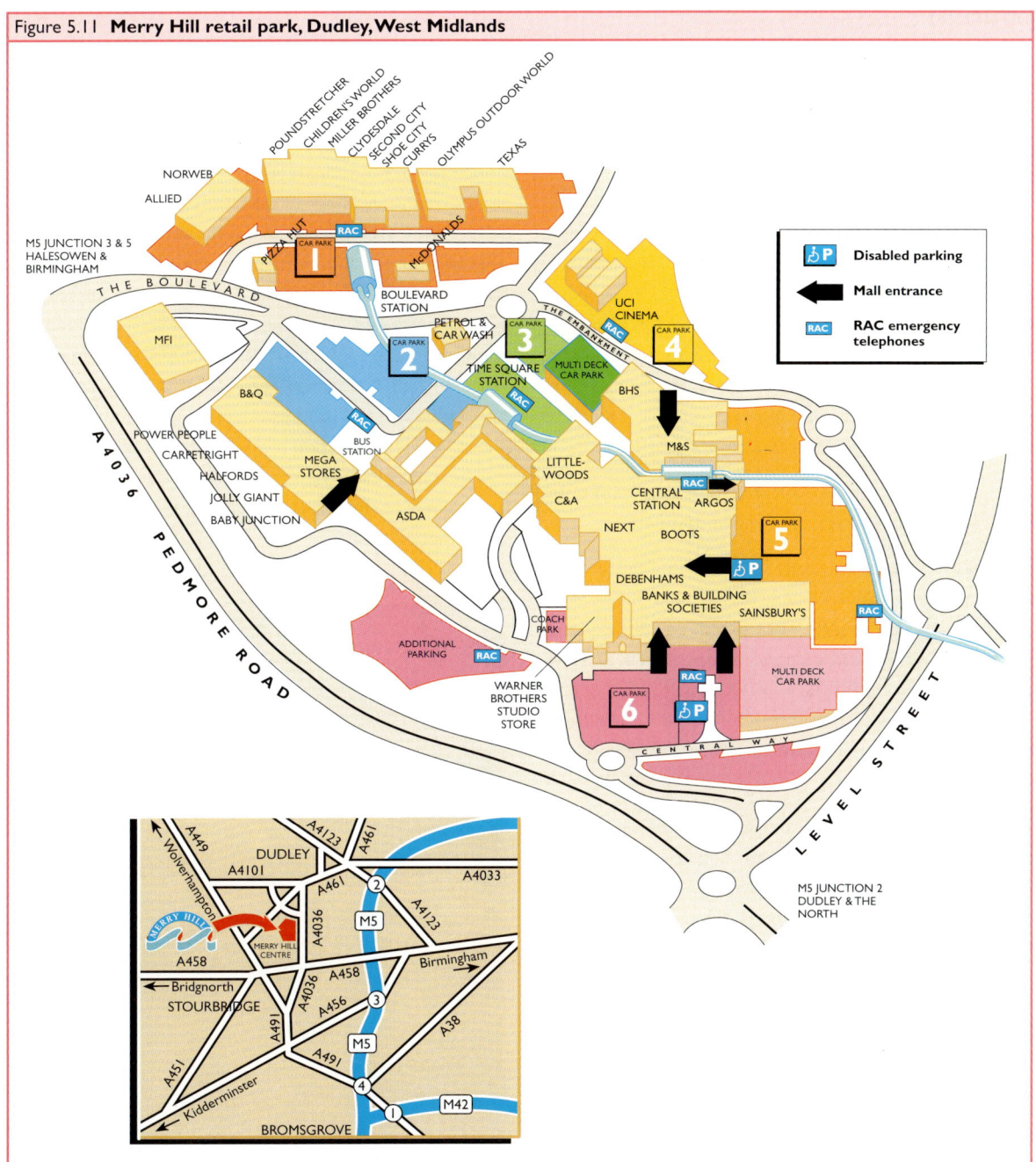

Figure 5.11 **Merry Hill retail park, Dudley, West Midlands**

Consequently these shops offer the **comparative advantage** of competitive prices combined with shopping under one roof.

Retailing **focus** refers to the need to match each outlet's location to the consumer group that is targeted. For example, the Sock Shop or Tie Rack is found in places with a large pedestrian flow, such as airports and major railway stations. New housing developments and renovation areas are extremely important for DIY operators while more exclusive companies and department stores prefer the high street locations and the controlled shopping malls. These all illustrate the locational requirements of retailing.

TOURISM

The growth in tourism in recent decades can be explained by a number of factors:

1. An **increase in disposable incomes** and **leisure time**.
2. **Improvements** in **transport and communications**.
3. Developments in **mass tourism** and packages holidays.
4. A broadening of lifestyle **expectations**.

Tourism is a vital industry in many countries, especially the more peripheral ones. It accounts for 21% of Spain's GDP, 5.1% of France's and 3.9% of the UK's and is a major source of foreign earnings. With the exception of Italy, tourism accounts for over 13% of the foreign earnings of Mediterranean and Alpine countries, compared with 4% for the countries of Northern Europe.

The impact of tourism varies throughout Europe. A number of areas stand out as major tourist destinations:

1. **Coastal resorts**, especially in the Mediterranean.
2. **Mountainous areas**, notably the Alps and the Pyrenees.
3. Areas of **scenic beauty**, such as Ireland and Scotland.
4. **Capital cities** and cultural centres, such as Paris, Oxford, and Bruges.
5. **Leisure parks** such as EuroDisney and Center Parcs.
6. **Business-conference** centres such as Korpilampi outside Helsinki.

In the UK, employment in tourism is concentrated in the South East, South West and East Anglia reflecting more affluent populations, a higher proportion of elderly people and a concentration of tourist potential, such as in London, Oxford and Cambridge. By contrast, lowest rates are found in Northern Ireland, owing in part to the Troubles, the Midlands and the North. These regions are characterised by low disposable incomes, above average rates of unemployment and poor images. Tourism in Northern Ireland boomed in the mid-1990s following the 1994 ceasefire and the hot summer of 1995.

Most countries have experienced increases in foreign tourists since 1960. Low cost package tourism associated with the growth of the major tour companies and the development of inclusive charter flights has been one of the most significant factors contributing to the rise of tourism in the Mediterranean. Although tourism is of increasing importance for employment, its impact is likely to be reduced for a number of reasons: strong dependence on female labour, unskilled, seasonal and part-time jobs and a movement towards self-catering accommodation.

There is a very complex pattern of tourist demand and provision throughout Europe. The type of person that is able to take holidays is influenced by a number of social and economic conditions: **stage in the family life-cycle**, **amount of leisure time available**, **access to tourist areas** and **disposable income** are all important. In general, manual workers, older people, large families and rural households are less likely to take holidays in another country. These conditions are more likely to be found in the less developed economies such as Italy, Spain, Greece and Portugal. By contrast, in the UK and Germany the proportion of people taking holidays shows very little variation related to class or age.

The tourist industry in Europe will continue to develop and change in the foreseeable future. Growth is linked to a number of trends:

Figure 5.12 **Tourism in Greece**

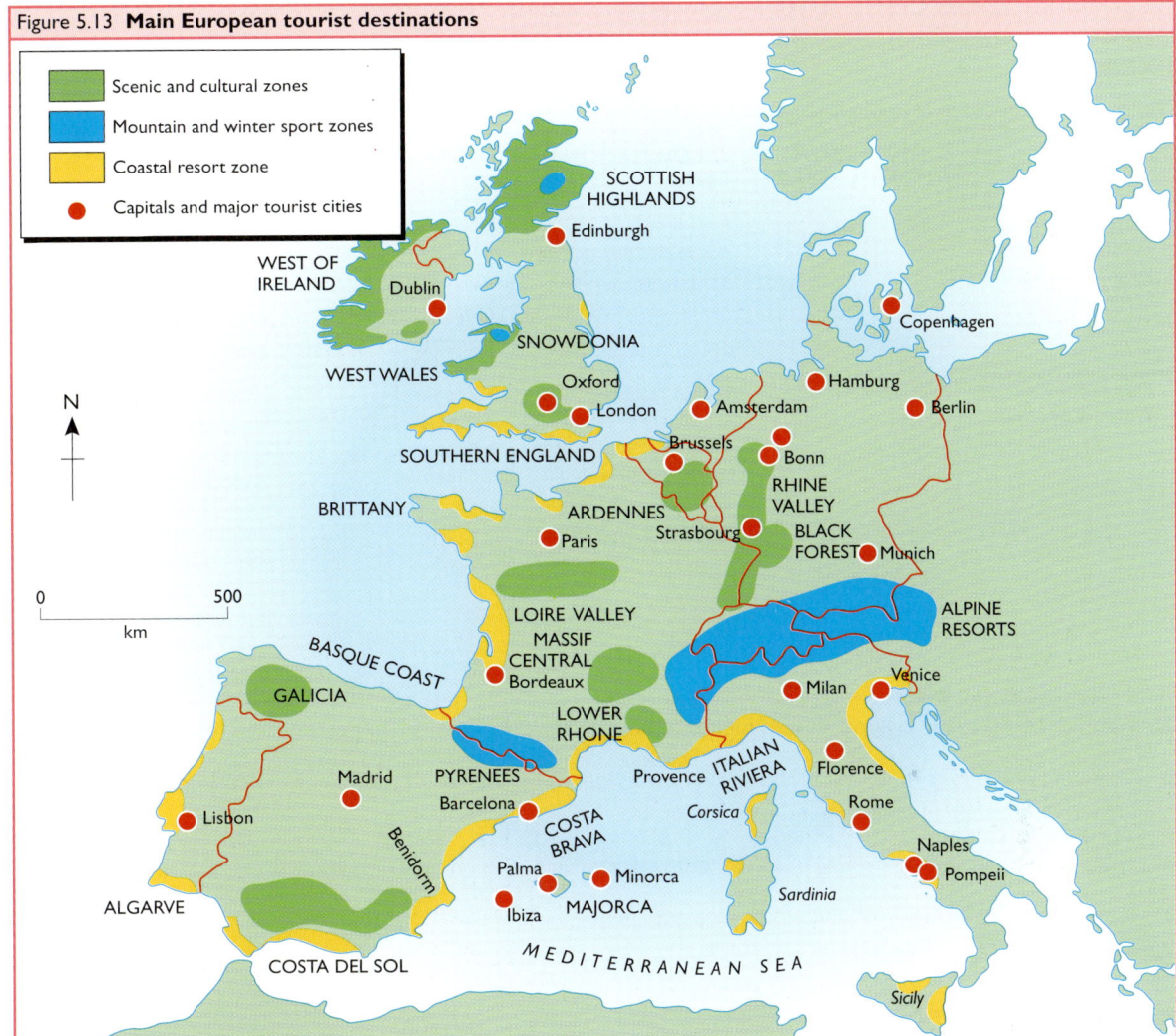

Figure 5.13 Main European tourist destinations

1. **Increased size of market** with tourists from the developing economies.
2. Growth of **second holidays**.
3. **Short breaks**, especially in urban locations.
4. **Quality and cultural tourism**.
5. **Overseas investment** such as USA in Greece.

Tourism has a role to play in regional economic development although certain factors need to be considered to assess its impact: is it integrated into the existing economy? does it utilise existing resources? is it being used to revitalise depressed areas? There are misgivings over its impact as it may create a dependency culture, caused by fluctuations in economic and political conditions and currency devaluations. Although the tourist industry is very labour-intensive, much of it is associated with part-time, seasonal and female labour. Similarly, the effect of tourists and tourism on the natural and cultural environmental will influence the nature of future developments. For example, rural areas have benefited from expansion of tourism, second homes and retirement populations. On the other hand, they have suffered disproportionately from the closure of schools, hospitals, post offices, shops and other services. Achieving the 'right' development for any area or region is a difficult task and requires appropriate consultation, planning and implementation.

Section C Exercises and recommended reading

EXERCISES

Figure 5.14 **Explanation of regional shift of high-technology firms**

LOCATION	% EMPLOYMENT INCREASE 1981-7	ORGANISATIONAL LEVEL	EXAMPLE	EXPLANATION
Outer Southern England (Home Counties and East Anglia)	+8000 (26%)	large firms with HQ, admin and research units small indigenous firms	Texas Instruments in Bedford IBM (R&D) in Hursley, Hants 'Cambridge phenomenon'	availability of qualified workers role of universities good communications (M4) space constraints in London
Assisted regions of Wales & Scotland	+7000 (21%)	branch production units set up by large firms often to serve European markets & avoid tariffs some upgrading with addition of R&D facilities	'Silicon Glen' IBM (assembly) in Greencock, Renfrewshire Bosch in South Wales	MNCs seeking European base policy financial incentives boosterism of Scottish and Welsh Development Agencies e.g. 80% of foreign-owned computer companies report that incentives are a major reason high unemployment, so low labour costs small firms as MNC spin-offs

1 **Figure 5.14** shows the regional shift of high-tech firms in the UK. Descibe, and give reasons for, this shift. [12]

2 Using the case studies in this chapter, explain how useful Weber's model of least cost location is in explaining the present distribution of high-tech industry in Europe. [13]

3 Using examples, evaluate the role of service industries in regional economic development. [13]

4 In what ways do the high-tech and service industries in peripheral regions differ from those in core regions. Give examples to explain your answer. [12]

RECOMMENDED READING

There are a number of books and articles that provide useful accounts of the electronics, high-tech and service industries. Castells and Hall (1994) *Technopoles of the World*, Routledge (European examples in Chapters 5 and 8 and Chapters 9 and 10). CEC (1993) *The Future of Industry in the Global Context*, Monitor. Dicken (1992) *Global Shift*, Paul Chapman Publishing (Chapter 10 gives the global perspective). European Commission (1994) *Competitiveness and Cohesion: trends in the regions*, Luxembourg. Howells and Wood (1993) *The Globalisation of Production and Technology*, Belhaven Press. Keeble, D. (1990) 'High-technology industry' in *Geography*, vol. 75, part 4, pp. 361-364. (Keeble's work is the basis for much of the current discussion on high-technology industries in the UK). Texts which deal with services include Clout et al. (1989) *Western Europe—a geographic perspective*, Townsend (1993) *Uneven Regional Change in Britain*, Daniels (1985) *Service Industries: a geographic appraisal* and (1993) *Service Industries in the World Economy*. Pinder's *Western Europe* has useful chapters by Shaw & Williams and Burt & Dawson. Porter's *Competitive Strategy* (1980) illustrates many of the changes in the retail sector. For high-tech industries, see 'The UK's changing manufacturing base' (1994) in *Geography Review*.

CHAPTER 6
THE CORE

A **WORLD CITIES**
   A new type of city...**100**
   European business locations...**102**

B **THE RANDSTAD**
   Agriculture..**106**
   Manufacturing industry..**108**
   Services..**109**
   Regional problems...**110**
   Planning solutions..**112**

C EXERCISES AND RECOMMENDED READING..**114**

Section A examines how world cities play a dominant role in the global space economy. The globalisation of industry, services, finance and government have played an important role in rise of world cities. World cities act as business locations where corporate headquarters, government and financial institutions and specialist producer services are located in close proximity to each other. World cities, such as London, Paris, New York and Tokyo are more likely to spread innovations between themselves rather than down the urban hierarchy. Hence the differences between world cities and other cities is likely to increase in the future.

World cities are growth centres. There is a concentration of economic, financial and political power. However, the image of these cities as centres of specialist producer services is somewhat limited. World cities have many problems: both London and Paris have experienced deindustrialisation, counterurbanisation, environmental dereliction, congestion, pollution and an array of social problems such as high rates of crime, ethnic tension and unemployment.

Section B analyses an urban-regional core, the Randstad. This covers an area of some 6,000 km², about the same size as London or Paris but it does not have the image of a world city. Functionally, the northern wing is dominated by the service sector while the southern wing is dominated by manufacturing and distribution. In the centre is the 'green heart', an area of intensive export-oriented farming. The area has long been the focus of the Dutch economy with heavy industries linked to processing imports, such as oil refining and chemicals, and light industries attracted by the urban markets. It is a core not only in terms of manufacturing industry, but also its agriculture and service sector are intensive by nature and export-orientated. The Randstad is centrally located between the major trade routes of Europe and it has a long tradition of trade and commerce. This is partly because the Netherlands has too few internal resources to support large urban economies, thus developments outside the national boundary have been crucial to the nation's prosperity. This competitive character combined with the shortage of space in the Randstad has created a complex set of problems that relate to agriculture, industry, urban development, transport and environmental protection.

Section A World cities

The future of large cities has been questioned. In the 1970s cities and their suburbs diverged. The dual processes of counterurbanisaton of people and decentralisation of manufacturing and retail establishments left cities with an identity crisis. The advantages of suburbs were more space, lower rents, better housing and schools, and cleaner air. The result was that the populations of London and Paris both dropped by nearly 20% during the 1970s.

However, in the 1980s cities and suburbs began to grow again. The boom in financial services and tourism has given a boost to downtowns. In the 1980s the population of Paris levelled out and that of London rose from 6.7 million in 1981 to around 7 million in 1991. Cities clearly have much to offer: a large pool of labour, large markets and a skilled and knowledgeable workforce.

A NEW TYPE OF CITY

There is a new kind of city emerging in the 1990s. Economic activities have shifted from the traditional pattern of mixed use to a **concentration on financial services**. Fruit and vegetable markets have moved away from Covent Garden and Paris Les Halles to avoid congestion. Manufacturing industries prefer accessible, out-of-town locations where greenfield sites allow purpose-built plants which are often low-rise and expensively landscaped. This process is symbolised by Renault's decision to move its factory from a site near the Eifel Tower to the suburbs (**Figure 6.1**).

Cities such as Paris and London are no longer centres for manufacturing. They are now the centres of **MNC headquarters**, **financial services** and **telecommunications**. They influence world affairs and their links to a few other similar cities like New York, Seoul and Rio de Janeiro has led to the concept of the world city being developed.

Figure 6.1 **Renault's car plant on an island in the Seine, Paris, which was moved to the suburbs**

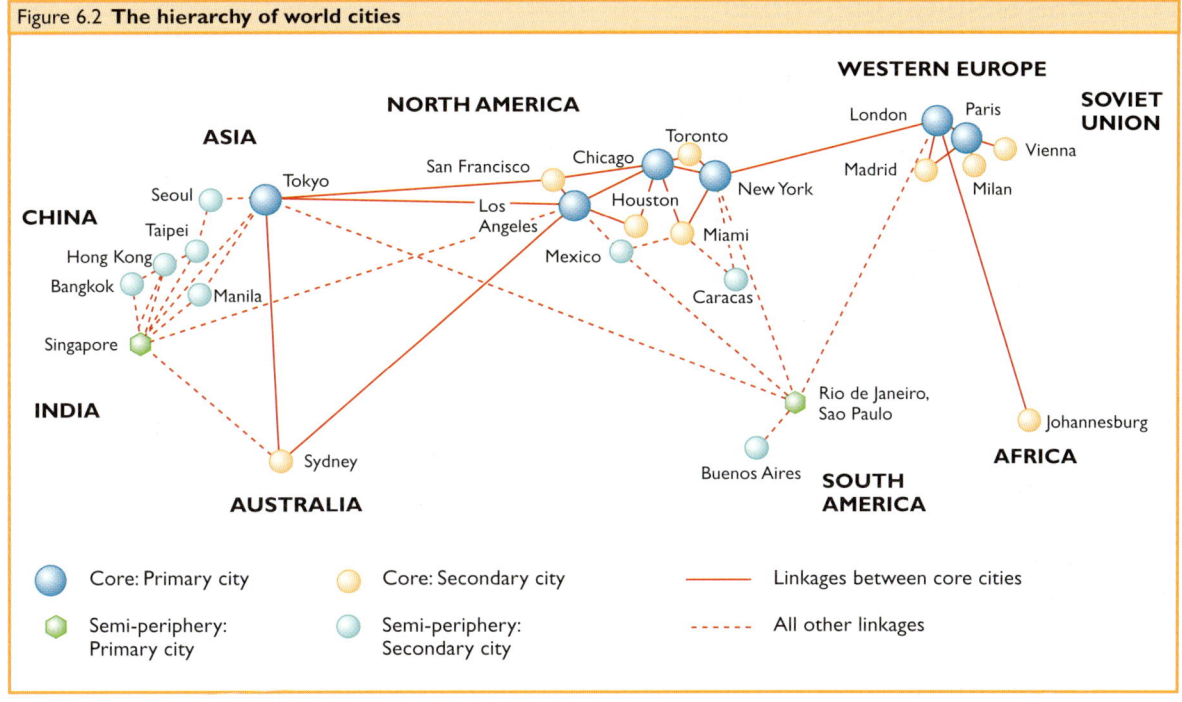

Figure 6.2 **The hierarchy of world cities**

Friedmann's world city

In 1986, John Friedmann developed the world city hypothesis. It was an attempt to understand the major global cities of the world and their response to the shift from an international to global economy. Friedmann has defined world cities as:

1. Centres through which money, information and commodities flow.
2. Large, urbanised regions defined by dense patterns of interaction.
3. Hierarchical with respect to the economic power they command and their ability to attract global investments.
4. Sites for the concentration and accumulation of capital.

Figure 6.2 shows Friedmann's hierarchy of world cities. European world cities are difficult to categorise because of their relatively small size and specialised functions. London and Paris are classed as world cities of the highest order but there is also the Randstad agglomeration (**pp.106-13**), and the German economy centred on Frankfurt.

The world cities have been created by the shift from an international to global economy. An **international economy** sees goods and services traded across national boundaries by individuals and firms from different countries under strict control of individual nations. However, in the **global economy** goods and services are produced by large MNCs who largely dictate the industrial policy of the nation state. They usually orchestrate their operations from world cities. The concept of the world city has been developed by other geographers. Appadurai (1990) has developed a list which describes the variety of functions which world cities offer. There are six cultural and economic landscapes:

1. **Ethnoscapes** produced by flows of business personnel, guestworkers, tourists, immigrants and refugees.
2. **Technoscapes.** Flows of machinery, technology and software from MNCs and government agencies.
3. **Finascapes.** Flows of capital currency and securities.
4. **Mediascapes.** Flows of images and information through newspapers, televisions and film.
5. **Ideoscapes.** Flows of a Western view of life, e.g. democracy, welfare rights and mass consumption.
6. **Commodityscapes.** Flows of culture and style encompassing everything from architecture and interior design.

The world city concept clearly has great relevance to any discussion of the location of businesses in European cities.

EUROPEAN BUSINESS LOCATIONS

Figure 6.3 shows the results of a survey based on interviews with 500 executives in nine EU countries. It takes into account 11 indicators, ranging from access to markets and telecommunications, to cost and availability of staff and accommodation.

World cities in Europe

London emerges as the favourite business location. It is, along with Tokyo and New York, one of the three centres of international finance. It also has a number of other advantages over its European rivals:

1. English is the accepted business language.
2. Value for money in office space.
3. Availability of relatively low cost labour.
4. Access to markets and links to much of the foreign investment in the EU.
5. Access to the Channel Tunnel.
6. The 1993 Finance Act which introduced a privileged tax regime for international headquarters (HQs).

Paris is second. It is one of the best-run big cities in the world. Unlike London, its mayor wields great power and the result is a very high level of public investment both in buildings and transport. In 1995 the annual budget of the city was Fr 30 billion (£4 billion).

Paris and **London** can be considered as Europe's only true world cities. However, other cities are also very successful. **Frankfurt** is a leading financial capital and also the most important city for the HQs of manufacturing industry. **Brussels** is the most important political centre, and **Amsterdam** with its links to **Rotterdam** is a major centre for industry and trade. The result is that more than 330 of Europe's 500 largest companies have HQs in just 10 cities.

Europe also has a degree of **specialisation** within its cities which is unusual. The linkages between each city and the freedom of movement allowed by the Single Market means that Europe's cities will fulfil different roles: capitals of finance, politics and manufacturing are emerging from the cities of the EU.

There is also a tendency to promote regions within Europe which span national boundaries.

Figure 6.3 **Best locations**

Rank 1994	City	Type of city
1	London	World city
2	Paris	World city
3	Frankfurt	European specialist city
4	Brussels	European specialist city
5	Amsterdam	European specialist city
6	Zurich	European specialist city
7	Barcelona	National specialist city
8	Düsseldorf	National specialist city
9	Milan	European specialist city
10	Madrid	Capital city
11	Manchester	National specialist city
	Munich	National specialist city
13	Berlin	European specialist city
	Stockholm	Capital city
15	Glasgow	National specialist city
	Hamburg	National specialist city
	Lisbon	Capital city
	Lyon	National specialist city

Figure 6.4 **Manchester's financial district**

Cities versus region

The EU's '**hot banana**' of most favoured regions is discussed in **Chapter 1**. It is a banana-shaped zone that sweeps from London, through Paris and Amsterdam, to Milan. It clearly does correspond to Europe's major cities but it is perhaps more aligned to the EU's **most favoured regions**: South East England, the Benelux countries, northern France, the Rhine and the Ruhr, and northern Italy. These regions are most likely to attract head offices and research facilities because of a number of locational advantages: market access, transport links, modern telecommunications and availability of labour.

The **Single Market** which lifted economic restrictions across Europe has seen other functions for cities. The following two case studies illustrate this. The Four Cities Partnership is an interesting example of the way cities can be used to promote regional links between European countries. The case study of Manchester illustrates the way cities can specialise in a national as well as European or global context.

Case Study 1: Manchester

One of the results of the boom in financial services in the 1980s was that the economy of the South became overheated. Cities such as Manchester and Leeds benefited from the high prices for land and services in London. A **spatial division of services** away from London allowed Manchester to become a self-sufficient regional financial capital. Banking, insurance, finance and business services now account for 111,200 jobs in the city. Manchester has seen an in-migration of lawyers from London. Salaries are similar but Manchester has lower living costs and does not suffer the problems of negative equity faced by workers in the South. There is also evidence of European linkages in Manchester's growth as a financial capital. The city is a member of the European Association of Regional Financial Centres which links it with Barcelona, Bilbao, Birmingham, Dublin, Edinburgh, Leeds, Lille, Lyons, Stuttgart, Turin and Valencia. All those cities show that successful regional specialisation is possible within Europe even given the dominance of world cities like London in the global economy.

Case Study 2: The Four Cities Partnership

The **Four Cities Partnership** (Figure 6.6) is a loose alliance between the cities of **Nottingham** in the UK, **Nancy** in France and **Karlsruhe** and **Halle** in Germany. Formed at the beginning of the 1990s it is based on an exchange of industrial expertise and joint research and business ventures. Each city has attempted to change its function in the light of industrial change:

1. Nottingham was hit by the recession of the early 1990s and has had to diversify its industrial base to include **services and business HQs**.
2. Karlsruhe also has a diverse economy based on **services and R&D**.
3. Nancy has moved away from heavy industry to concentrate on **services and academic institutions**.
4. Halle is emerging from high unemployment due to **deindustrialisaton** of its machine tool and chemical industries.

The result is an extension of the policy of **twinning towns**. There is an exchange of expertise such as the advice Karlsruhe has given Nottingham on developing its rapid transit system. Nottingham has also taken on some ideas pioneered by its sisters in the Four Cities. It has set up a 'technology region' offering low cost sites to design companies modelled on a similar scheme in Karlsruhe. For Nancy twinning with Karlsruhe has attracted not only the usual student and sporting exchanges but also provided a powerful ally in lobbying for a Franco-German rail link. For Halle the partnership has been a chance to develop away from its linkages to the old East Germany. Karlsruhe is obviously the dominant member.

The Four Cities Partnership is an interesting venture. Such cross-regional linkages do provide a role for Europe's medium-sized cities. It remains to be seen whether this type of venture can develop the business or educational links which will promote economic growth or whether they represent a more cosmetic '**Europeanisation**' of cities which have failed to find a context in the urban hierarchy of their own countries.

Figure 6.5 **A new office development in Nottingham**

Figure 6.6 **The Four Cities Partnership**

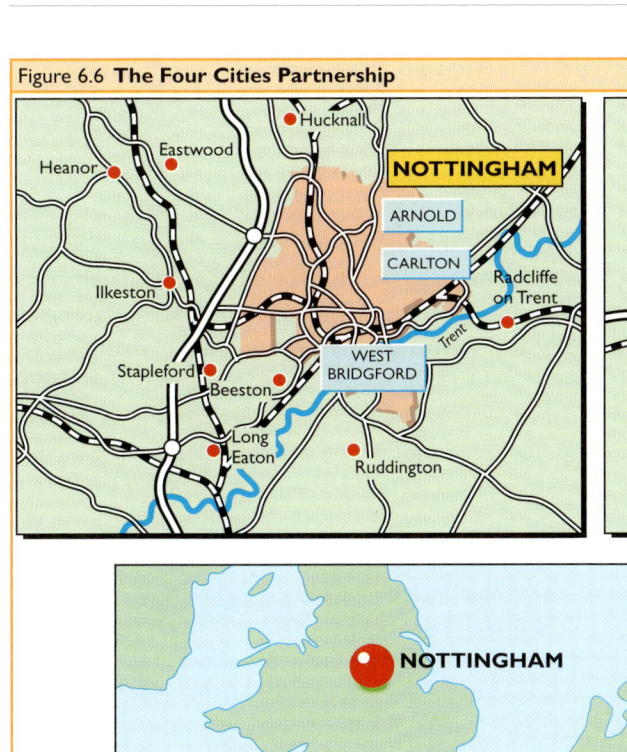
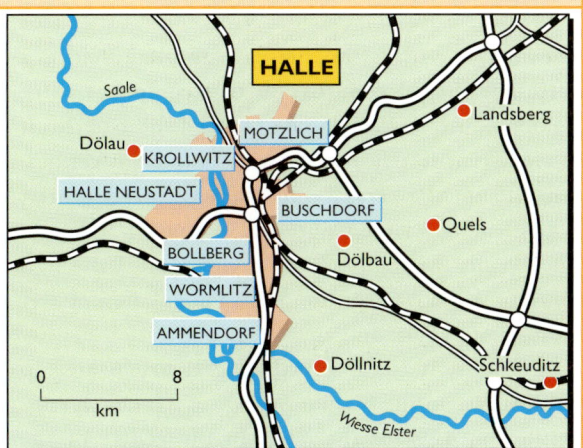

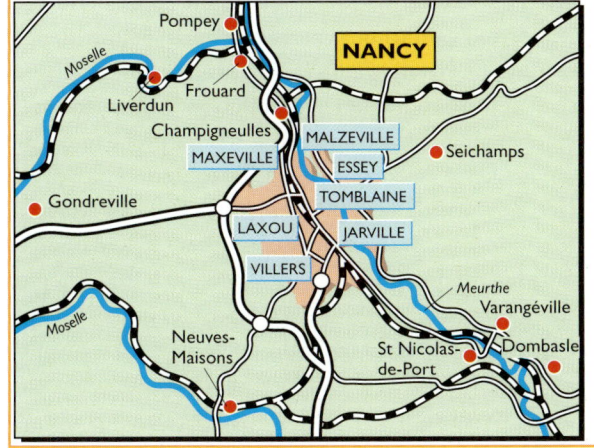

Section B **The Randstad**

Known as the 'ring-' or 'rim-city', the Randstad is a horseshoe-shaped conurbation that comprises three regions, North Holland, South Holland and Utrecht. It is a polycentric city region, with four main centres: Amsterdam, the capital and main financial centre; Rotterdam, the largest port and centre for industry; The Hague, the centre of government; and Utrecht, a conference and trade centre. With a population of over six million people, and an average population density of 450/km², it is the fourth largest urban area in Europe, accounting for almost 43% of the Dutch population on 16% of the land. In the centre of the Randstad there is an open agricultural area known as the 'green heart', an area of intense market gardening, dairying and horticulture accounting for 40% of the Netherlands' agricultural income. The Randstad's economy is generally quite strong and stable, with a broad base and a heavy dependence on foreign trade. Despite high labour costs and taxes, the Netherlands has attracted 20% of American and 33% of British investment to the EU since 1987, mostly to the Randstad.

AGRICULTURE

Although agriculture is not normally associated with core areas, in the Randstad it is **high value** and **capital intensive** (high-tech farming). Within the green heart there are some of the most intensively farmed areas of Europe: **bulb growing** at Aalsmeer; intensive **dairying** west of Utrecht; **market gardening** under glass in the Utrecht area; and intensive **animal production** in the east. Agriculture is **small-scale** but **highly efficient** and **specialised** and there is a growing trend towards larger and even more mechanised farms. Dutch horticulture, centred on the Randstad, accounts for 60% of world trade in cut flowers and the agricultural sector accounts for 5% of employment and 25% of total exports, typifying the country's export orientated economy. Nevertheless, recession in the flowers' auctions' major markets meant that after a number of prosperous years sales fell in the mid-1990s.

A number of factors explain the highly intensive nature of Dutch agriculture. There is a **shortage of land** and 40% of farms are less than 10 ha, and nearly one-third of the land is reclaimed polderland, and thus productivity must be high to repay the cost of this land and make a profit. **Competition** for land from the rapidly growing Randstad cities creates small holdings. To be successful farming must be intensive and most farmers concentrate on **high quality** agricultural goods such as dairy products, horticulture and market gardening. There is a favourable **physical environment**: low-lying relief, coastal dunes protect the crops from westerly winds, and a **moderate climate** influenced by oceanic conditions gives an early spring, reduces the danger of frost and gives an early start to the growing season. The **soil** is a mixture of fertile alluvium in major river basins and light sandy soils in the higher heathlands of the east and south east, which respond well to fertiliser and deep ploughing.

Human factors also favour Dutch agriculture. Farmers are well **educated** and **trained** in modern farm practices and have access to capital for mechanisation, fertiliser, irrigation and drainage control, and quality animal breeding. The **centrality** of the Randstad is excellent: there is a large nearby market of over 6 million people and good transport links with the major markets of Europe. This allows for the **rapid distribution** of high value perishable goods. Moreover, Dutch natural gas provides a relatively cheap source of heating for greenhouses and the Dutch chemical industry provides a wide range of relatively cheap fertilisers.

However, farming is not without its **problems**. The green heart is under renewed threat from urban, industrial and even agricultural development. The growth of greenhouses **destroys the beauty of the area**, and there is widespread **pollution**. The Netherlands' Fourth Report on Physical Planning (1988) stressed environmental issues such as **overproduction** and **pollution** by intensive agriculture: ammonia pollution is a serious problem on the sandy soils of the south and east. By contrast, much of the land on the **reclaimed polders**, which is not as attractive for industrial or residential purposes, is left unused. **Surplus production** in the EU, and the introduction of the quota system, has reduced the need for intensive farming on new land. As a result the polders, which were initiated at a time of land and food shortage, and were very expensive to create, cannot be used for farming, but are given over to **nature reserves**.

Figure 6.7 **The Randstad and agriculture**

MANUFACTURING INDUSTRY

Manufacturing, like agriculture, is not usually connected with the urban core. However, in the case of the Randstad manufacturing is concentrated in the south, in particular with the port of Rotterdam (**Figure 6.8**). The success of Dutch manufacturing lies in its **export orientation**, **restructuring**, diversity and **economies of scale**. Although not long established industrially, the Randstad contains a substantial proportion of the Netherlands industrial population. It accounts for a major share in the Dutch economy: 32% of GDP, 25% of employment and investment in the Netherlands, and 30% of the country's gross exports. Although large, its relative share of the Netherlands' industrial output is not as great as it is for agriculture or for services (**Figure 6.9**). It is largely dependent on the foreign market both for raw materials and sales. Major exports include **gas**, **organic chemicals**, **artificial resins and plastics**, **agricultural products and foodstuffs**, **machinery** and **electronics**. About 30% of Dutch exports go to Germany although increasing amounts are directed to developing countries. Other important service industries include the gas and chemical industries (Shell, Unilever and Gist Brocades), the transport equipment and metal industries (Chubb Lips and Fokker), the manufacture of instrumentation, the food, drink and tobacco industry (Heineken) and building construction. A number of oil companies operate giant storage tanks near Rotterdam and the region plays a crucial role in Europe in the landing, storage and refining of oil products. Large-scale heavy manufacturing is concentrated around Rotterdam and the North Sea canal, e.g. metals at Ijmuiden, foodstuffs at Zaandam, and Hoogovens, a large steel and aluminium works on the North Sea coast employing about 2,500 workers.

Change in manufacturing

During the 1980s the industrial sector experienced **deindustrialisation** and **decentralisation**. Employment fell from 40% in 1950 to 14% in 1991. Traditional sectors such as metal works, textiles and the food industry contracted severely. **Reindustrialisation** occurred through the expansion of information technology, chemicals and precision engineering. However, growth in export orientated industries, such as metal and chemicals, was restricted by competition from Eastern Europe. The decline in manufacturing was not as severe as elsewhere in Europe, since the Netherlands has never been associated with '**Fordist**'-type employment but has depended more on trade and distribution.

Centrality in Western Europe is the key factor in understanding the success of the port of Rotterdam. It is the world's largest seaport, supported by a vast transport and distribution network. Not only is there large-scale industrial development but there are enterprises in other sectors, associated with international transport, such as banking, insurance and commerce. Its dependency on **trade** and entrepot function for other industrial areas in Europe has necessitated a highly developed transport system, characterised by a dense network of roads, rail, waterways, and oil and gas pipelines. Most of Rotterdam's growth has occurred since 1945. A number of factors help explain its role as the world's leading port, in addition to its centrality: transport links, large-scale manufacturing developments, its role as an entrepot for oil imports, and government incentives to expand Rotterdam's varied roles.

Figure 6.8 **Rotterdam: gateway to Europe**

SERVICES

The service sector in the Randstad has a long tradition. The 'golden age' of trade with the Far East in the seventeenth century has been replaced by a wide range of services centred on Amsterdam. The service sector includes the **business and financial sectors**, and also the **transport sector**. In the Netherlands these are especially important given the **export orientation** of the economy: 23% of all Netherlands' transport activities are concentrated in the Randstad.

The Dutch economy is typified by export, in terms of trade and number of firms and the presence of foreign firms. The Randstad offers an **accessible site** for **head offices** and **distributive centres** for international companies in particular and the Netherlands has actively attracted Far Eastern firms to Amsterdam: Nissan, Sony, and IBM have major plants near Schiphol and the North Sea canal and almost half of the Japanese firms located in the Netherlands are found in the Amsterdam district.

While employment in agriculture and manufacturing industry is declining, employment in services is increasing. From the mid-1980s, employment in business services increased 34%, and banking and insurance 16%. However, there have also been large job losses in the transport equipment industry and the food, drink and tobacco industry.

Utrecht

Utrecht's employment is not only characterised by service employment, but also by the large number of high ranking jobs, e.g. managerial and professional employment. About 20% of these jobs are concentrated in Utrecht. Service employment is dominated by the **head offices** of large banks such as AMRO, insurance companies such as AMEV, the Jaarbeurs congress centre, information firms and over 500 international firms including IBM and Digital.

Nevertheless, Utrecht's service sector is dominated by **small firms**, with 85% of firms having fewer than ten employees. It has a relatively large **information-intensive** service sector, an equally large transport and distribution sector and a small labour intensive industrial sector. The number of jobs in the service sector increased by 30% in the late 1980s and early 1990s, a rate above the national average. However, the growth has tended to be in the outer areas of metropolitan Utrecht rather than in the city itself and about one-third of jobs are occupied by women. The working population has increased considerably since the early 1980s and is higher than the national average, owing to the younger age structure, the high proportion of working women and the high proportion of the working population with advanced educational and vocational qualifications.

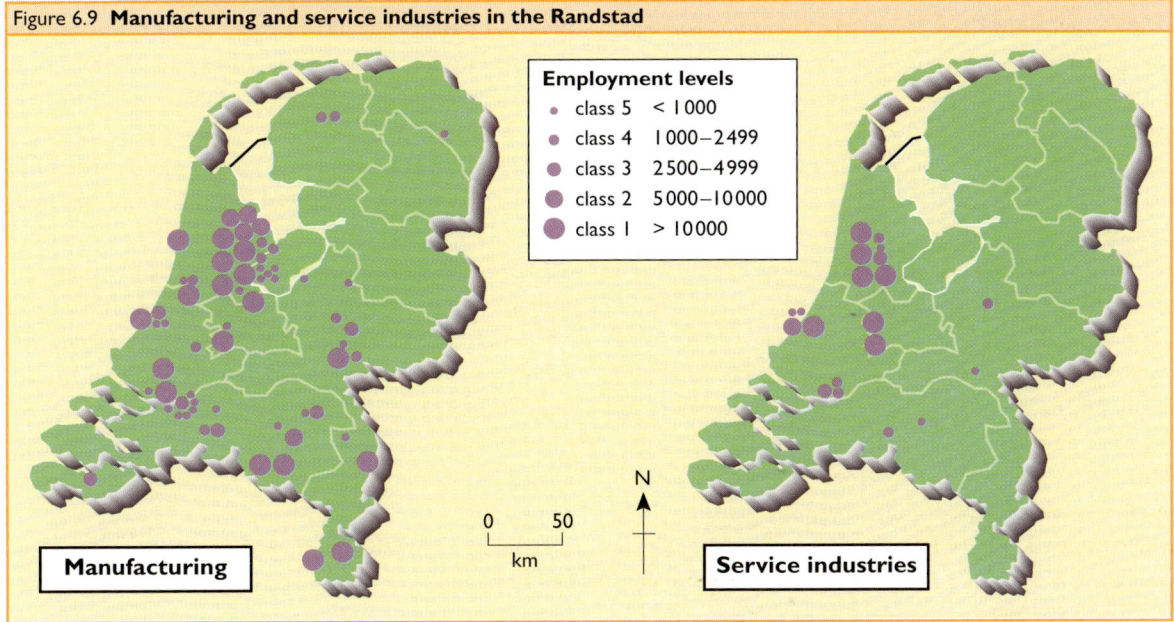

Figure 6.9 **Manufacturing and service industries in the Randstad**

REGIONAL PROBLEMS

The Randstad is a thriving region. It is one of the core regions of Europe yet it is not without problems. These stem from the limited land availability and the demands of a prosperous people and an expanding economy. Pressures arise from five interrelated factors: environment, demography, agriculture, industry and transport.

Urban sprawl and **suburbanisation** has created serious pressure on the green heart. Amsterdam, Rotterdam and The Hague lost 25% of their population between 1965 and 1985. There is continued demand for low density housing (6 houses/ha) but only a limited amount of very expensive land. The Dutch show a preference for single-family homes in less congested rural areas leading to large scale **suburbanisation** and **counterurbanisation**, especially by young families. During the 1970s the population of Amsterdam, The Hague and Rotterdam declined by 350,000. Since 1960 1.5 million people have been catered for with 600,000 homes, using up 100,000 acres of good quality land (**Figure 6.10**). As a result of the land price the cost of housing is two to three times greater than in Eastern Holland. Moreover, uncontrolled urban growth could create a European **megalopolis** including the Randstad, Rhine-Ruhr, and Brussels-Antwerp-Ghent.

Population pressures have created a significant shortage of resources. Growth has been rapid, from 0.5 million in 1800 to 1 million in 1900 and to over 6 million in 1995. Population densities are over 1,000/km^2 in some places. The large proportion of immigrants, 'frontier commuters' and migrant workers from Europe, and their associated high birth rates and levels of unemployment have created much social tension. A social polarisation between the wealthier, middle class residents of the newer suburbs and the lower income, elderly, ethnic minorities, overconcentrated in run-down, inner cities developed into a major social and political issue.

Since the 1960s **population growth** has been less than the national average because overspill has occurred into neighbouring Flevoland. There has been an effect on population structure: the population is ageing and household composition is changing. Since 1960 the proportion of the population aged under twenty has fallen 16% while that aged over sixty-five has risen 72%. More single people are living alone and there are more single-parent families, especially in the urban areas. This ageing has

Figure 6.10 **Green heart: under threat?**

been offset by the large numbers of immigrants into North Holland, approximately a third of all migrants to the Netherlands. The foreign proportion of the Randstad's population, mostly Turks and Morrocans, is 6%, concentrated in Amsterdam, 24%, Rotterdam, 9%, and The Hague, 8%.

Unemployment in the Randstad is mostly **structural** and has been relatively steady at about 4% since the mid-1980s. The widespread restructuring of Dutch industry has benefited service industries, in particular transport and distributive trades. The long-term unemployed (over one year) form a high proportion of the total unemployed, 40% in South Holland. Many of the unemployed are under-educated and belong to ethnic minorities and are frequently concentrated in inner city areas. The general increase in employment opportunities does not benefit them.

Industrial growth is putting severe pressure on the limited land that is available, especially in the green heart, the coastal strip and around deep-water ports. Rotterdam, for

Figure 6.11 Congestion in Amsterdam: roads are heavily used—only water transport moves efficiently

example, grows at an annual rate of 6,000 ha. Many of the former industrial sites are contaminated and in need of cleansing. The large-scale concentration of petrochemical works around Rotterdam is increasing the problem of acid rain. Moreover, the Randstad suffers as a result of industrial activity in the Ruhr polluting the Rhine.

Transport and communications expansions create a significant threat, Schiphol Airport in particular. The close proximity of the airport to large urban areas is an environmental hazard, as well as a potential threat to human life (**Figure 6.12**). The shortage of land has become intensified by competing demands for land-uses. The Netherlands is a **gateway to Europe** and the transport infrastructure is crucial to the success of the Dutch economy. However, only recently has it developed a coherent transport policy. The key element is to strengthen the two main ports, Rotterdam and Schiphol, as the gateway to Europe. Consequently, the Randstad has been late in developing its motorway network. Before the 1970s there was no motorway linking the principal cities. Commuting remains a major issue in the Randstad and as many as 2 million cars converge on Amsterdam and Rotterdam daily.

Figure 6.12 Schipol air-crash, October 1992

PLANNING SOLUTIONS

Although the Randstad is a core region its numerous problems require a regional policy. However, the close association of the Randstad and the national economy, and the small size of the Netherlands, has enabled planning to succeed at a national level, if not necessarily at a regional one. Planning in the Netherlands has at various times been dominated by social (1950s and 1960s), economic (1970s and 1980s) and environmental concerns (1990s) and the measures adopted have included at least three types of approach:

1. A **Christallerian** approach based on **central place theory** with a hierarchy of settlements providing the maximum service with the least number of outlets, e.g. Zuider Zee, South and East Flevoland polders, with new towns at Lelystad (1967) and Almere (1976) designed to accommodate 250,000 and 100,000 respectively.

2. A **growth pole approach** based on **Perroux's** model whereby a settlement with the best prospect for growth is targeted and developed as a regional centre for innovation and development, e.g. Alphen and Gouda.

3. **Growth axes**, as used in Paris, whereby linear chains of settlements are identified as having potential for further growth and development, e.g. between Breda to Eindhoven in the south east of the region.

Dutch urban planning has adopted elements of all three approaches.

Planning has had a strong grip on urban development since World War II. **Early policies** dictated that no Dutch city should have more than one million inhabitants. Population dispersal was stimulated by providing incentives to problem regions and decentralising government functions. Altogether the planners attempted to prevent existing settlements from coalescing, preserve the green heart and guide urban growth (**Figure 6.13**). The **Second Report** developed the idea of 'concentrated deconcentration', similar to the expanded towns policy in the UK. and two new towns, Lelystad and Almere, were built on the reclaimed polders.

During the 1950s and 1960s the Randstad's population increased more rapidly than the rest of the Netherlands. Traffic **congestion** and worries about the loss of agricultural land in the green heart led to a policy of dispersal beyond the green heart and there was also strict control over building there. **'Buffer zones'** or green wedges were established between the main towns to prevent them from merging and using up agricultural land and overspill zones were designated directing outwards along four principal growth.

New towns were built for a variety of reasons: to accommodate overspill populations, reduce pressure on the green heart, create employment, provide a more pleasant environment to live in, provide an administrative role and reduce the pressure on major cities. The main weakness of this central place approach was inefficiency, as resources were spread too thinly. However, during the 1970s problems of **decentralisation** became more serious and the inner cities experienced population losses: out-migration led to a reduction of 500,000 people, 11% of the total population. By contrast, the total population of the area as a whole increased 3.5% p.a. and built-up areas 7% p.a. over this period. The increase in commuting led to congestion in the cities but there was also a reduced tax base for the cities as industries moved to greenfield sites. The inner cities were characterised by low income groups and ethnic and social segregation. This continued during the 1980s: the population of the four largest cities in the Randstad declined by 15% whereas the fringe areas and the green heart increased by 67.1% and 53.8% respectively. Thus there were renewed fears that **large-scale dispersal** of employment and population would weaken the economic base of the Randstad, which is of fundamental importance to the continued prosperity of the Netherlands. Consequently, **reverse commuting** has increased: the unemployed of the inner cities have to travel to the suburbs and outer metropolitan areas in search of employment. This causes a great strain on the transport infrastructure and creates associated pollution problems.

Decentralisation was replaced by **regeneration** of inner cities by reducing traffic congestion, building new houses, renovating old housing, controlling suburban sprawl and improving environmental conditions. This 'balanced development' proposed that 35,000 new dwellings be built each year and a similar number renovated in inner city parts of the Randstad. Disused warehouses and buildings around the inner docks in

Figure 6.13 **Overspill, growth areas and urban intersections, Randstad**

Amsterdam and Rotterdam were to be turned into modern apartments. **Overspill** from the major cities was directed to and concentrated in a small number of 'growth towns' in, or near to, the Randstad, e.g. Alphen and Gouda. The concentration of resources in a small number of growth points is more likely to generate multiplier effects than the preceding central place approach. The large-scale development of the 1960s was scaled down. The remaining green heart will be vigorously conserved for agricultural and recreational use.

By the 1990s planners favoured '**urban intersections**' or nodal growth points. These were selected on account of their location in the transport network, economic activities and function as regional service centres. A threefold classification emerged (**Figure 6.13**):

1 **International centres**: e.g. Amsterdam, Rotterdam and The Hague.
2 **National centres** based on growth poles, e.g. Utrecht, Eindhoven.
3 **Regional centres** based on service centres, e.g. Maastricht.

Similarly **environmental concerns** and **traffic congestion** were perceived as major problems in the Randstad. Environmental issues have had considerable implications on housing development, e.g. high-rise blocks of flats have been replaced by low density dwellings of 45-50 units/ha.

Nevertheless, despite the environmental movement, economic development is crucial for the success of the Randstad. Important elements include the growth of Rotterdam and Schipol as gateways to Europe, the attraction of high-tech producer services, Dutch agriculture and the urgent need to update the transport infrastructure. Balancing these conflicting land-uses will continue to torment planners in the foreseeable future.

Section D Exercises and recommended reading

EXERCISES

1 **Figure 6.14** gives data on London's inputs and wastes.
   (a) Why are London's inputs not equal to its outputs? [8]
   (b) Why is input of water so large? [8]
   (c) How do you think London's status as a world city contributes to its problems. [9]

Figure 6.14 **London's inputs and wastes**			
INPUTS	tonnes/year	**WASTE**	tonnes/year
Water	1,002,000,000	Inert	8,200,000
Oxygen	40,000,000	Industrial	3,000,000
Fuel (oil equivalent)	15,900,000	Household	2,470,000
Food	2,400,000	Digested sewage	7,500,000
Phosphate	72,000	CO_2	60,000,000
Timber	1,200,000	SO_2	400,000
Tropical hardwoods	200,000	NO_x	280,000
Paper	2,200,000		
Plastics	2,100,000		
Glass containers	210,000		
Plate glass	150,000		
Cement	1,940,000		
Bricks, sand, tarmac	6,000,000		
Metals	1,200,000		

2 **Figure 6.15**, taken from *Geographical*, June 1994, suggests a sustainable city system which could be implemented in any large city. What are the advantages and disadvantages of such a scheme? [25]

Figure 6.15 **Implementing a sustainable city system**
To achieve greater sustainability a substantial proportion of the waste outputs of cities should be turned into their inputs in the following ways:
• Sewage works would become fertiliser factories for enriching farmland
• Washing powders, toilet cleaners and bleaches would be fully biodegradable
• Companies would invest in recycling technology
• Organic household waste would be recycled for compost
• Industrial and household waste would be burnt to produce heat
• Forests would be replanted not only for timber but also for protecting watersheds and absorbing carbon dioxide

RECOMMENDED READING

The world city hypothesis is discussed in depth in 'World Cities in a World System' (1995) edited by Paul Knox and Peter Taylor. The book also reproduces Friedmann's original article 'The World City Hypothesis' from 1986. Other good sources are Castells and Hall (1994) *Technopoles of the world*, de Smidt, Granberg and Wever (Eds) (1991) *Regional development strategies and territorial production complexes—a Dutch-USSR perspective* and Dieleman and Musterd (1992) *The Randstad: a research and policy laboratory*.

CHAPTER 7
REGIONS OF CHANGE

A **THE UK'S NORTH-SOUTH DIVIDE**
- The industrial mosaic 116
- Northern restructuring 118
- The overheated South 119
- The regional future of the UK 119

B **THE WEST MIDLANDS**
- Background 120
- Inward investment 121
- Regional policy 122

C **ITALY AND REGIONAL DEVELOPMENT**
- The North-South divide in Italy 122
- 'Third Italy' 126

D **THE RUHR**
- Regional policy 128
- Environmental industries 129
- Technology transfer centres 130
- The future 130

E EXERCISES AND RECOMMENDED READING 132

This chapter discusses the strategies which regions and states can employ when faced with industrial change. 'Change' usually means a decline in one economic activity and the search for opportunities in another. A vast array of policies can be used to stimulate regional growth but the individuality of Europe's regions means that a standard approach is impossible.

Section A looks at the UK's North-South divide. In many ways the British government's regional policy aims to encourage inward investment by foreign MNCs to replace its own old-fashioned manufacturing firms. In the 1980s MNCs flocked to Britain's regions and the North saw a resurgence in its manufacturing base and a relocation of service industries from London. **Section B** uses the West Midlands as an example of a region which embraces elements of both the South and North of England. The example of Jaguar is used to illustrate the effects of inward investment on regional development.

Section C discusses regional policy in Italy. It goes beyond a simple evaluation of the Mezzogiorno, Europe's classic problem region. It argues that there are three distinctive regions in Italy and while growth pole strategy has not been successful in the South, the flexible specialisation which has had good results in the 'Third Italy' is not necessarily suitable throughout the country.

Section D centres on another of Europe's classic problem regions, the Ruhr. Here a decline in coal and iron and steel has seen a change in the industrial base with a move towards environmental industries and also a policy of technology transfer.

Section A The UK's North-South divide

The recession of the early 1990s modified the traditional North-South divide in the UK.

For the first time, the key service industries of the South were hit by the economic downturn. Consequently, the South suffered more from the recession than the North, something Britain had not experienced before. One explanation is that the North emerged from the recession of the early 1980s lean and hungry for manufacturing growth, while the South grew complacent on its seemingly infallible service sector. Figure 7.1 reveals the different industrial profiles of Britain's regions. In the West and East Midlands, over 20% of the workforce is engaged in manufacturing. By contrast, well over 50% of employees in Greater London and the South East work in services. In 1991, a third of Greater London's gross domestic product was generated by financial and business services, compared with only a tenth in the North.

THE INDUSTRIAL MOSAIC

The very different industrial mosaic of the UK can also be seen in **Figure 7.2**. It shows that at the broadest level there exists a **North-South divide** in terms of a split between services in the South and manufacturing in the North. As a result, the last recession has affected the UK's regions in different ways:

1. At a national scale, **business failures** in 1992 were nearly a third higher than in 1991, with the largest rise in the South East region outside London (46%), followed by the South West region (35%) and Scotland (36%).

2. Between 1990 and 1992, all regions showed a **reduction in the number of people employed in manufacturing**. This was most marked in the South West, where more than one in six jobs were lost, and in Greater London, where the job loss was one in seven.

3. During the same period, in the **financial services** sector around 20% of employees in East Anglia and 10% in the South West lost their jobs (this sector grew by 32% in Wales in the two years to June 1992).

Figure 7.3 shows that the North escaped the worst of Britain's rise in unemployment during the early 1990s. Unemployment has risen fastest in the South East, the South West and East Anglia. Between 1990 and January 1993 the change in seasonally adjusted unemployment rates was less than 2% in Northern Ireland, the North and Scotland, but around 6.5% in the South West.

To understand this apparent convergence of regional fortunes it is necessary to look back to the cycle of recession which until 1989 had led to a widening of the North-South divide.

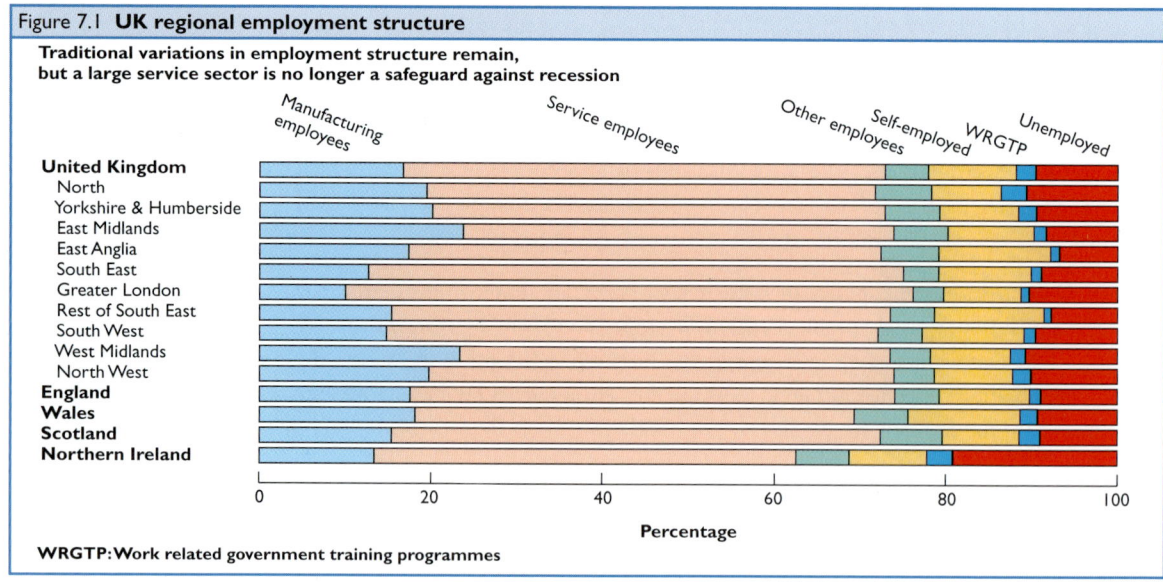

Figure 7.1 **UK regional employment structure**

Traditional variations in employment structure remain, but a large service sector is no longer a safeguard against recession

WRGTP: Work related government training programmes

THE UK'S NORTH-SOUTH DIVIDE 117

Figure 7.2 **The North-South divide and the new map of regional policy**

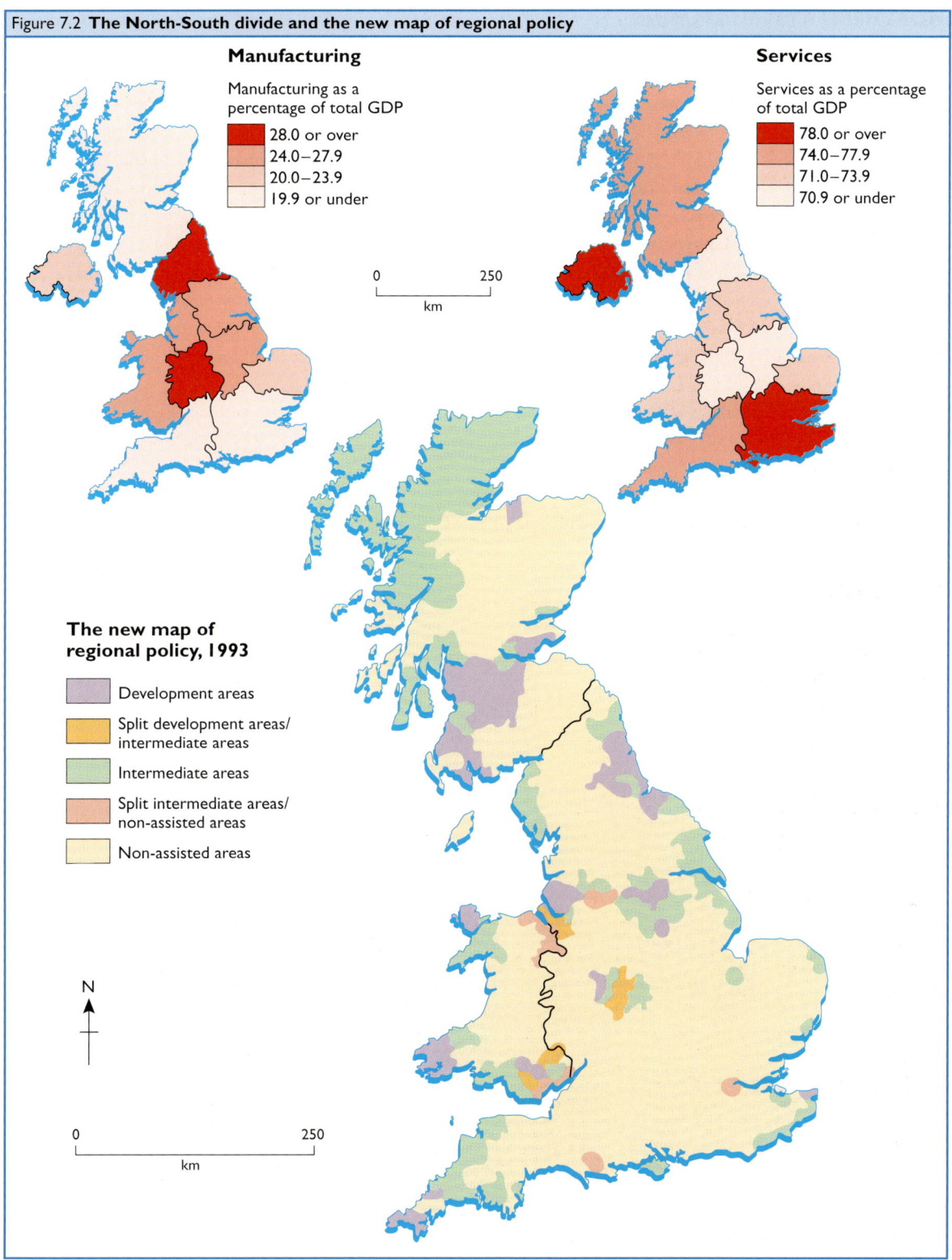

NORTHERN RESTRUCTURING

The Northern regions are emerging from the recession with a diversified and restructured economic base and with few of the problems of debt, negative equity and services decline suffered by the South. They have emerged blinking after a very dark period in the early 1980s when the recession hit Northern manufacturers very hard and left the South largely untouched. There followed a painful period of **redundancies** and **rationalisation** which led to the collapse of many businesses; but the survivors emerged strengthened.

At the same time, foreign-based **MNCs** were attracted to the North by aggressive regional pro-growth coalitions promising cheap labour and land in addition to the grants available under the terms of assisted area status. Fired by the fear of the **Single European Market**, MNCs crowded through the UK's open doors during the 1980s. Over the past 40 years, 38% of US and 39% of Japanese investment in the European Union has come to the UK. In 1990, some 21% of investment, 19% of output and 13% of employment in UK manufacturing was accounted for by foreign-owned companies. The regional statistics are even more telling. The North has attracted three times as much **inward investment** as the South with Northern Ireland earning more than 33% of its manufacturing revenue from MNCs compared to less than 15% in the South West.

Figure 7.3 **Unemployment in the UK**

Percentages	1990	1991	1992	January 1993
United Kingdom	5.8	8.1	9.7	10.6
North	8.7	10.4	11.3	12.1
Yorkshire & Humberside	6.7	8.4	9.9	10.6
East Midlands	5.1	7.2	8.9	9.7
East Anglia	3.7	5.8	7.6	8.6
South East	4.0	7.0	9.4	10.5
Greater London	5.1	8.2	10.6	11.7
Rest of South East	3.1	6.0	8.3	9.5
South West	4.4	7.1	9.1	10.0
West Midlands	5.9	8.6	10.7	11.5
North West	7.7	9.4	10.5	10.9
England	5.3	7.8	9.7	10.6
Wales	6.6	8.7	9.7	10.3
Scotland	8.1	8.7	9.5	9.9
Northern Ireland	13.4	13.8	14.5	14.7

Claiming unemployment benefit, seasonally adjusted annual averages.

THE OVERHEATED SOUTH

The 1990s saw the **Southern economy overheat** as service industries expanded and a rise in house prices led to an unsustainable cycle of borrowing and spending. Interest rates were pushed up to dampen spending and the housing market collapsed. House prices in the South fell on average by more than a quarter between 1993 and 1995. 40% of those who bought property in Greater London after 1987 have mortgages exceeding the value of their homes, i.e. **negative equity**. And yet the problems facing the South can only partly be blamed on the property slump. The South began the 1990s with a number of disadvantages:

1 The South **lost industries** in the mid to late 1980s because of the high land prices, congestion and labour shortages caused by a mini-boom in the region.

2 The long run of success in the service sector bred **complacency**. While the North was attracting inward investment through its boisterous development agencies, the South was actively discouraging it: anti-growth coalitions were formed to protect the quality of life in the leafy Home Countries.

3 At the same time the **decline of the defence industry**, due to the post-Glasnost peace dividends, led to the removal of the £8-9 billion worth of 'hidden' regional aid to the Home Counties in the from of government defence research and development expenditure.

The skills factor

The South is no longer fending off new inward investment and moves are afoot to compete with the development agencies of the North.

With respect to employment, the South has one major comparative advantage over the North, above and beyond its geographical position. It is the skills of its workforce, much of which stems form the concentration of the university-educated middle classes in the Home Counties.

The South East and South West also have the highest proportions of 16 year olds staying on at school or going on to further education, over 75% in 1991-2. The quality of the South's labour represents a substantial regional advantage and this disparity is perhaps the North's greatest obstacle in building on its resurgence in the 1990s.

THE REGIONAL FUTURE OF THE UK

It is clear that some of the Northern regions are in a strong position. **Leeds** and **Manchester** have diversified into **regional service centres**. **South Wales** outperforms the rest of the UK when it comes to **valued added** per employee in manufacturing. **Scotland** seems to be developing an industrial base centred on its new towns and has attracted some **key inward investment** projects: most notable was Hoover's decision to rationalise by closing its French plant at Dijon and concentrating its operation in Glasgow.

Furthermore, the influx of overseas investment has stimulated the northern economy through demands for local goods and services and also through its spread of diverse management techniques. The North must now work to keep hold of its MNCs. Foreign industries were attracted to the UK and specifically the North because of low domestic costs. A resurgence in the economy could lead to a rise in labour costs and relocation away from Britain. Regional development agencies must now develop an aftercare and support role to ensure that the investors regions have gained remain and prosper. Indeed, future growth may well depend on expansions and reinvestments of existing MNCs.

A number of futures are possible for the UK's regional economy:

1 An **export-led recovery** would favour regions like South Wales, the North West, Yorkshire-Humberside and the Edinburgh-Glasgow axis. These areas have benefited most from inward investment and a restructured manufacturing base.

2 The prospect of a South Eastern recovery could be prevented by its poor manufacturing base and the negative equity problems of many of its workforce, leading to a further **redistribution of wealth and economic activity to the North.**

3 The growth of the North may be uneven, leading to a **greater polarisation** in the UK's regional geography and a decline in outer peripheral areas like the Highlands, Northern Ireland and western Wales.

4 There could be a **shift back to the dominance of the South** as a result of its superior skills base, its proximity to the EU's growth poles, a more aggressive approach to attracting inward investment and the assisted area status of selected coastal towns.

Section B The West Midlands

The West Midlands is one of the most interesting regions in the UK with respect to economic geography. It can be viewed as one of the poorer regions in the EU, with a low skills base and inner city areas with some of the most serious problems in the country. However, it can also be viewed as the manufacturing heartland of Britain, an area which has attracted substantial inward investment (**Figure 7.4**) and with competitive and low cost industries as result of recent rationalisation. It is also unclear whether the region belongs in the North or South of Britain. Traditionally, the West Midlands is the first region to enter a recession but also the first to emerge from one.

BACKGROUND

The West Midlands comprises the counties of Shropshire, Staffordshire, Warwickshire and Hereford and Worcester (**Figure 7.5**). It also contains the metropolitan areas of Dudley, Sandwell, Walsall, Rugby and Wolverhampton. The West Midlands Conurbation (Birmingham and the Black Country), Coventry and the North Staffordshire Conurbation represent the region's major urban areas. Historically the shires are rural-based and the towns have a variety of industries. The region has done much to redraw itself as a modern manufacturing and service centre. Organisations such as the **Black Country Development Corporation** have created new jobs and modernised the often derelict areas abandoned by older industries such as steel.

A number of factors influence the economic geography of the West Midlands: the increasing importance of **services**, the influence of **inward investment**, and **regional policy**.

Figure 7.6 shows evidence of a shift towards services. Employment in manufacturing fell by 430,000 jobs between 1979 and 1992 whereas jobs in services grew by 180,000. Coventry and Birmingham have established themselves as service centres of regional and national importance.

The boom in the late 1980s saw many financial services leave London because of high land and labour costs. The West Midlands with its good rail and motorway connections was a natural target. For example, Barclays Bank relocated six of its head office functions to Coventry along with 400 senior managers. A further 20 staff were transferred to Northampton. In a similar move the TSB moved its head office to Birmingham.

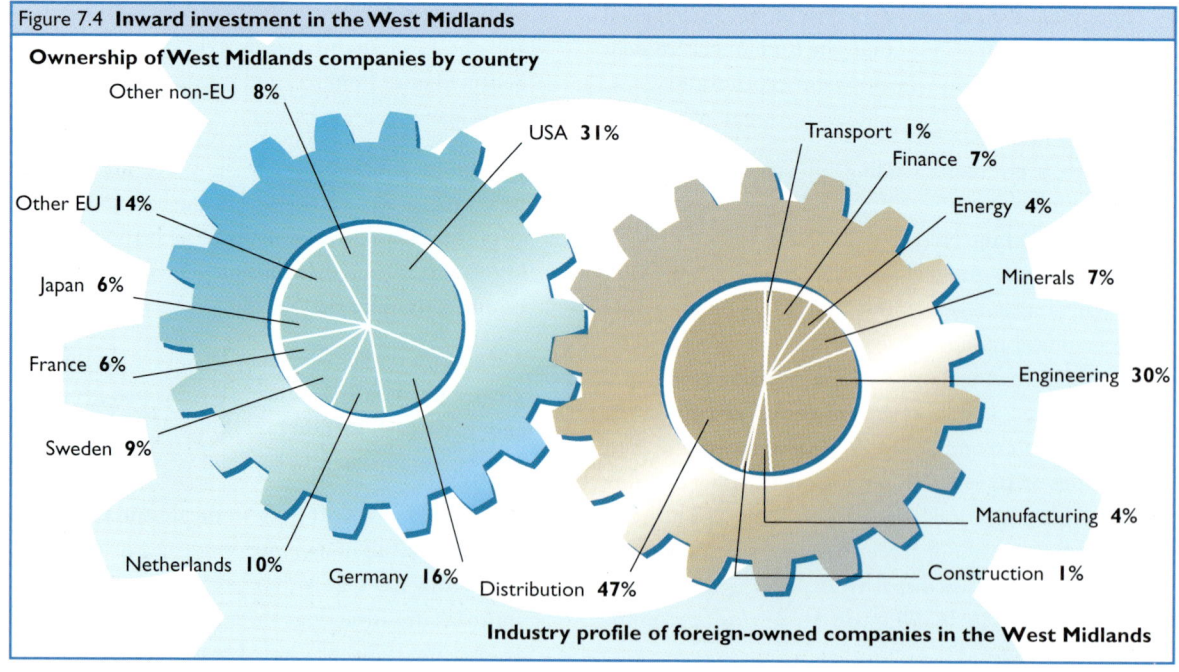

Figure 7.4 Inward investment in the West Midlands

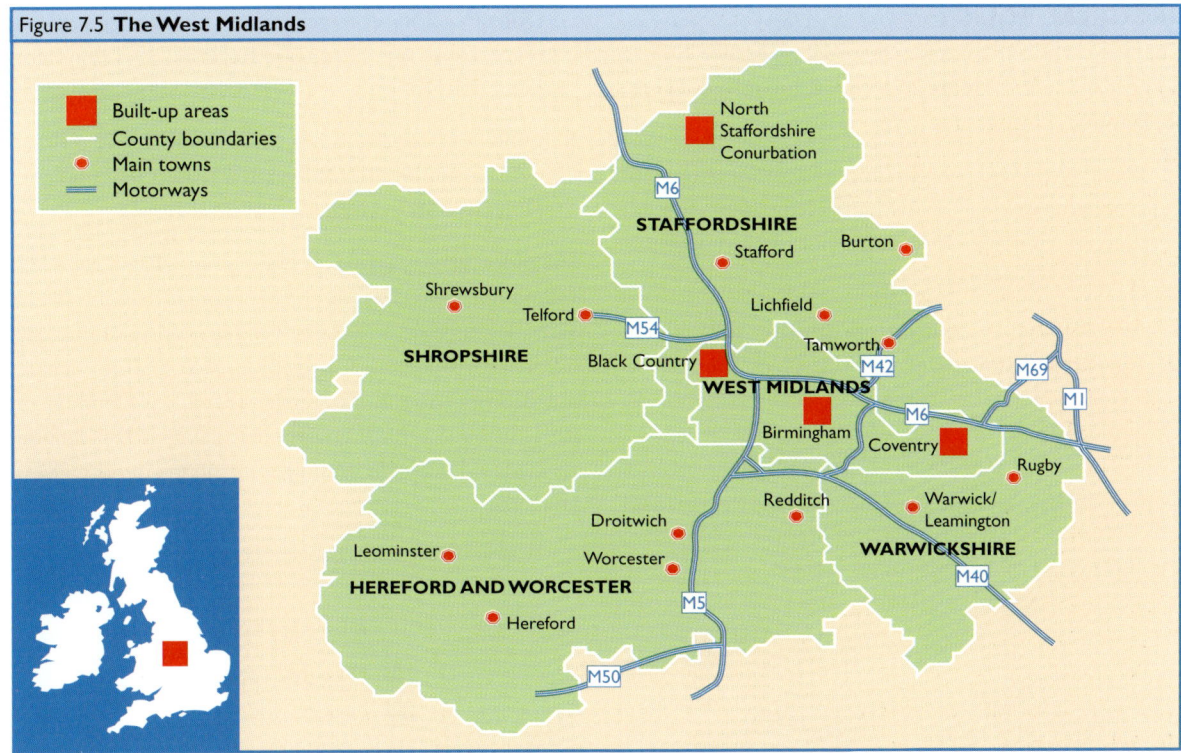

Figure 7.5 **The West Midlands**

INWARD INVESTMENT

The expansion of the West Midlands during the 1990s is due to a recovery in **engineering and the motor industry** on which the region is heavily dependent. In Birmingham, the Black Country, Solihull and Coventry, 40% of male jobs and 16% of female jobs rely directly on manufacturing. Much of this employment is from overseas firms who have invested in the region. Foreign investment accounts for about 20% of the UK total and since 1988 foreign investment in the West Midlands has topped £2.5 billion, creating 22,000 jobs. The area has been particularly successful in attracting **re-investments** for company expansions. Unemployment in the second half of the 1980s came down faster in the West Midlands than in the UK as a whole because of inward investment.

There is a growing concentration of **electronics companies** around Telford and the Warwickshire Science Park. However, the car industry dominates the region's manufacturing. Three factors account for this:

1. The expansion of car assembly in the UK.
2. The need for access to European markets.
3. The West Midlands is a cheaper location than many other areas of the UK—its labour costs are only slightly above those of Spain and half of those of Germany.

The region faces some locational disadvantages. It has a **lack of the greenfield sites** needed by many industries. There is also a reluctance of local manufacturers to invest in new plant and machinery. An even greater shortcoming is the poor skills base of the West Midland's labour force.

One of the most important locational characteristics of the West Midlands is its eligibility for regional assistance both from UK and EU sources.

Figure 7.6 **Changing employment structure, 1981-92**			
	Birmingham	West Midlands	Great Britain
Manufacturing	-43.5%	31.0%	-22.1%
Services	6.2%	16.1%	13.1%
Self-employment	37.8%	39.4%	34.6%
Total	-11.2%	-0.7%	5.0%

REGIONAL POLICY

The West Midlands receives financial assistance from a number of sources (**Figure** 7.7). In UK urban and regional planning:

1 Birmingham and Wolverhampton have development area status and their own urban development corporations which use government money to encourage private sector investment.

2 An east-west corridor from Rugby through to Coventry, Birmingham and the Black Country is classed as an intermediate assisted area.

In addition, funds are available from the EU:

1 In 1988 large areas of the West Midlands were granted **Objective 2** status (regeneration) which triggers assistance from the European Regional Development Fund (ERDF).

2 Other areas receive funds through **Objective 3** (long term unemployed), **Objective 4** (youth training) and **Objective 5b** (regeneration of rural areas).

The older, urban section of the region will receive £287 million between 1994 and 1997 through its Objective 2 status. Jaguar Cars is the largest single recipient of the UK aid. The two plants—Castle Bromwich in Birmingham and Browns Lane in Coventry—shared £9.4 million in 1994. In 1995 a huge government grant of £80 million persuaded Jaguar to upgrade its plants in the West Midlands.

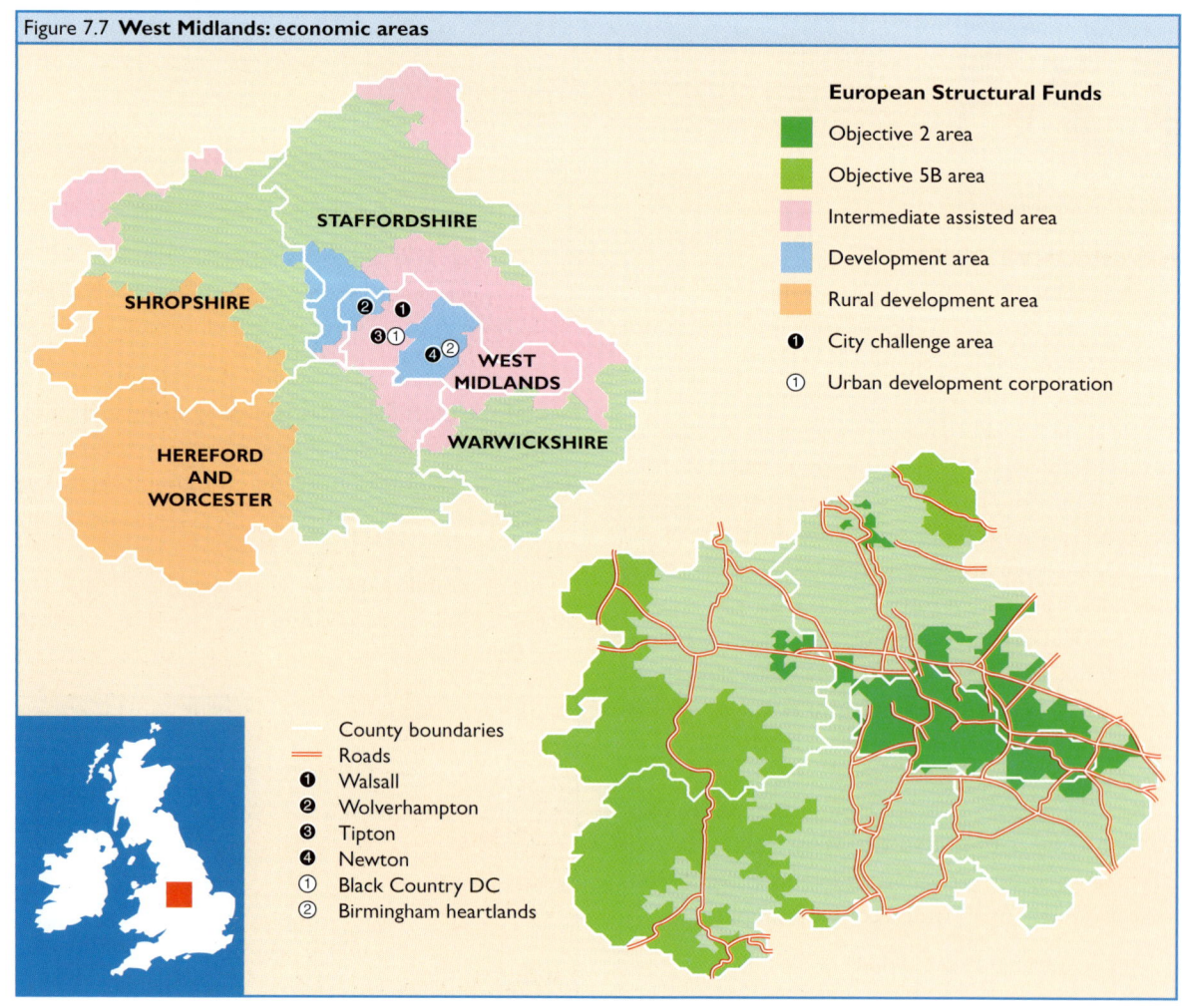

Figure 7.7 **West Midlands: economic areas**

Case Study: Investing in Jaguar

In July 1995 Ford announced a £400 million investment in Jaguar which will expand the Castle Bromwich plant in Birmingham (**Figure 7.8**). The decision will create 1,300 jobs by 1999 when the car is scheduled for production. Another 150 jobs will be created at Ford's Halewood plant on Merseyside which will be responsible for body pressings. The investment, which could have gone to Ford's plant in Detroit, was made possible by a generous aid package from national and local government (**Inset 7.1**). The money was spent to persuade Ford to invest. It shows a continued belief that large manufacturing plants can have a **multiplier effect** in stimulating the local, regional and national economy.

Ford's decision adds to the concentration of the car industry in the West Midlands. Rover, Land Rover and LDV have plants around Birmingham and Jaguar and Peugeot Talbot are long established in Coventry.

The investment will also stimulate the components industry. An estimated dozen components manufacturers will be drawn to the area creating around 5,000 jobs. Several of the firms are American companies eager to cross the Atlantic and retain their close links with Ford. For example, Lear Seating, a major supplier for Ford models in America, is producing almost all the seats for Jaguar cars on a just-in-time basis. Jaguar needs to be re-supplied with seats every 90 minutes and so close proximity is essential. The plant is located five miles away from the main Brown's Lane complex in Coventry.

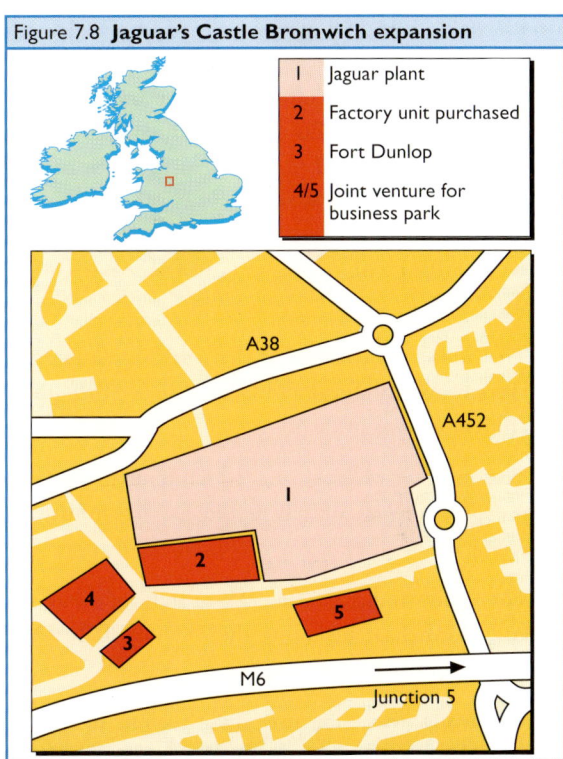

Figure 7.8 Jaguar's Castle Bromwich expansion

1	Jaguar plant
2	Factory unit purchased
3	Fort Dunlop
4/5	Joint venture for business park

Figure 7.9 Jaguar production line

INSET 7.1 THE AID PACKAGE

1. £48 million from the Department of Trade and Industry as regional selective assistance.

2. £15 million from English Partnerships in Property to ease the further expansion of the Castle Bromwich plant.

3. £17 million from local publicly funded agencies to provide for services such as training and environmental improvement.

Section C Italy and regional development

Italy is one of the most culturally diverse and regionally divided countries in the EU. In terms of national growth Italy has recently overtaken Britain as the fifth biggest industrial power of the West. However, this expansion has not benefited the whole country. Three clear regions can be identified with respect to industrial development (Figure 7.11):

1. The 'industrial triangle' of the North including the cities of Milan, Turin and Genoa—one of the most affluent regions of the EU.

2. The South or Mezzogiorno, which remains the least prosperous region of Italy despite massive influxes of regional aid.

3. The rapidly industrialising 'Third Italy' of the centre and North East including the regions of Tuscany, Veneto and Emilia-Romagna.

The following discussion uses these regions to discuss the causes of, and responses to, the inequal regional development within Italy.

THE NORTH–SOUTH DIVIDE IN ITALY

Northern Italy has been compared with Japan and Taiwan in its rapid and sophisticated industrial expansion. The cities of Milan, Turin and Genoa have proved particularly resilient to recession and have expanded their textiles, machine tools, leather goods, footwear and fashions firms which now sell to the rest of Europe.

The Ancona Wall

The South of Italy is one of the classic peripheral problem regions of the EU. The eight regions of the Mezzogiorno make up an area which has 40% of Italy's land area and 35% of its people but only 24% of national GDP. In Italy, the **'Ancona Wall'** is a popular term used to separate the prosperous North from the problem South. It extends south west from Ancona and divides Tuscany, Umbria and Northern Italy from the South.

Four factors have been suggested to explain the different rates of development between the North and the Mezzogiorno:

1. The **'Southernist view'** explains regional contrasts in political terms. The argument is that the South was neglected by Northern politicians who concentrated expenditure on public works and technical education in the North.

2. The **'two nations'** explanation points to fundamental cultural differences as the reason. A number of dichotomies illustrate the perception of the North as modern and the South as backward: efficiency versus inefficiency; urban versus rural; metropolitan versus provincial; entrepreneurial versus peasant.

3. The spatial explanation points to the South's **distance from the core** markets of the EU. The relative inaccessibility of the South is not helped by its poor infrastructural links to the North.

4. The **agricultural explanation** points to **rural decline** as one of the reasons for the poverty of the South. The harsh Mediterranean climate with long periods of intense heat and little rain made non-irrigated farming difficult. Absentee landlords who rented land to low paid peasants had little incentive to improve the land. This led to the migration of the most able workforce to the North or outside Italy to prosperous countries such as Switzerland.

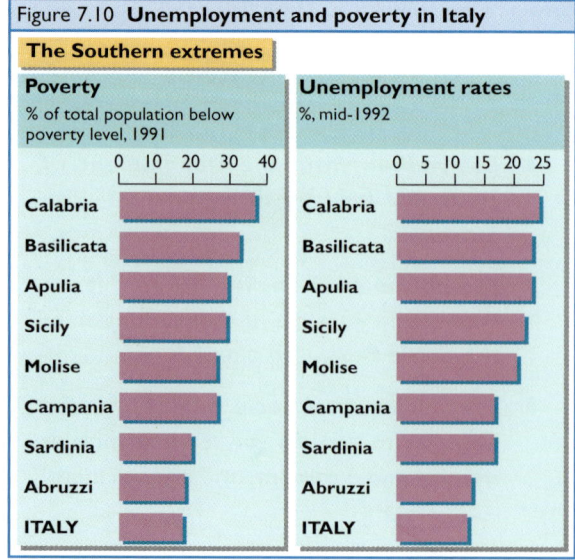

Figure 7.10 Unemployment and poverty in Italy

The policy

In 1950 the **Cassa per il Mezzogiorno** (Fund for the South) was set up to raise Southern living standards to those of the North. The Cassa aimed to do this in two main ways:

- **Land reform** which broke up the large semi-feudal estates of the South, creating about 120,000 new, small farms.

- A **growth pole strategy** which legislated that 60% of all state investment went to the South with investments in large steel and chemical plants—it was hoped the growth pole would stimulate the Southern economy attracting new capital, stimulating local firms and providing employment.

Enormous sums of public money were spent in the South during the period 1950 to 1980. However, the success in achieving the stated aim of providing a self-sustaining industrial and service base has been questioned. The big industrial ventures known as 'cathedrals in the desert' failed to stimulate the region and became 'museums in the desert'. They failed because:

- they were **capital intensive ventures** built without reference to the needs of the local economy or the markets for steel and petrochemicals which were in the North or outside Italy;

- the **Opec price-hike** which quadrupled the **price of oil** in 1973 raised the price of their raw materials;

- a process of **deindustrialisation** is under way in the South with the steel plants at Naples and Taranto in a process of rationalisation due to overcapacity in the EU.

The results

There are a number of viewpoints about the overall success of the Cassa. The generally held belief is that the South has become increasingly subsidised, dependent and incapable of self-generating growth. However, some parts of the Mezzogiorno are prospering. The regions of Abruzzi, Molise and Apulia along the Adriatic Coast are doing well and towns like Pescara, Taranto and Siracusa have grown rapidly in terms of population and prosperity.

However, the overall failure of Italian regional policy, especially the growth pole strategy, was recognised in 1986 when a law was passed to encourage private enterprise in

Figure 7.11 **The three Italys**

the South. Private investment was encouraged by a treasure trove of grants, loans and tax incentives, which effectively increased the financial burden on the state.

Fiat's plant in Melfi is a LIRE 4,600 billion (£1.8 billion) investment which has resulted from this policy. It represents a decision based on generous government grants and subsidies, lower land costs than in the North and the availability of greenfield sites. Indeed, Fiat, with automotive plants in Southern locations that include Avellino, Pomigliano near Naples and Cassino, now produces more from plants in the South than the North.

Even Fiat's Melfi plant can do little to dispel the conclusion that after over 40 years of regional policy the Mezzogiorno has been left with a **dual industrial structure** of Northern or foreign-financed giants (chemicals, petrochemicals, metallurgy, cars) and small artisan-style concerns (furniture, food stuffs, clothing) with little or none of the linkages predicted by the growth pole strategy.

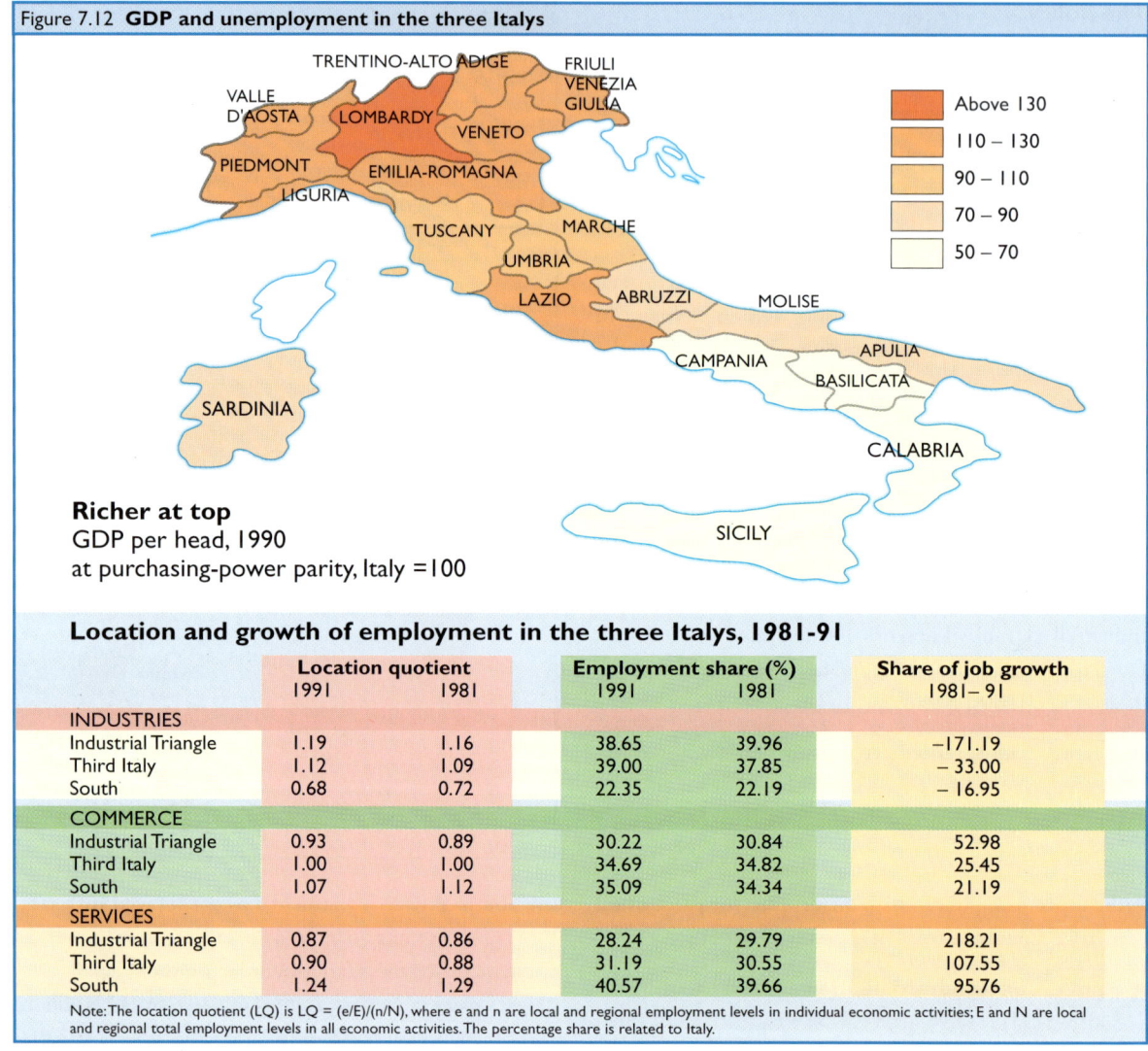

Figure 7.12 **GDP and unemployment in the three Italys**

Richer at top
GDP per head, 1990
at purchasing-power parity, Italy =100

Location and growth of employment in the three Italys, 1981-91					
	Location quotient		**Employment share (%)**		**Share of job growth**
	1991	1981	1991	1981	1981–91
INDUSTRIES					
Industrial Triangle	1.19	1.16	38.65	39.96	−171.19
Third Italy	1.12	1.09	39.00	37.85	− 33.00
South	0.68	0.72	22.35	22.19	− 16.95
COMMERCE					
Industrial Triangle	0.93	0.89	30.22	30.84	52.98
Third Italy	1.00	1.00	34.69	34.82	25.45
South	1.07	1.12	35.09	34.34	21.19
SERVICES					
Industrial Triangle	0.87	0.86	28.24	29.79	218.21
Third Italy	0.90	0.88	31.19	30.55	107.55
South	1.24	1.29	40.57	39.66	95.76

Note: The location quotient (LQ) is LQ = (e/E)/(n/N), where e and n are local and regional employment levels in individual economic activities; E and N are local and regional total employment levels in all economic activities. The percentage share is related to Italy.

'THIRD ITALY'

The Italian regions of Emilia-Romagna, Tuscany and Veneto represent an increasingly prosperous area between North and South known as 'Third Italy'. The success of Third Italy is based on a policy which is very different to the growth pole strategy attempted in the Mezzogiorno. Third Italy uses **industrial districts** composed of small firms to pursue a strategy of **flexible specialisation** based on close cooperation between firms, sub-contracting and links between factory and client. These areas also benefit from a strong tradition of **local enterprise**, high levels of education and good communications without the high land prices of Milan and Genoa.

Case Study: Textiles in Tuscany

Since Tuscany was targeted as an Objective 2 area to reverse industrial decline, its development has made it a European Union success story.

Within Europe Tuscany can be classed as a **peripherally ascendent region**. In 1970, Tuscany was 13 points below the EU mean in terms of GDP per head, by 1990 it was eight points above. Indeed, the performance of Tuscany can be compared with some of the most dynamic regions in France (e.g. Rhone-Alps and Champagne-Ardenne), Netherlands (north and south Holland) and the UK (the South East).

The region's textile industry based in Prato has benefited

from **European integration** and **regional decentralisation** but also from the industrial district approach to development. The city of **Prato** and a group of nearby centres in Tuscany constitute one of the three clusters of the **Italian woollen industry**. A process of change begun in the 1950s led to the creation of the Prato industrial district or Prato textile district. By 1991 Prato's textile industry employed 36,000 workers in 7,000 firms; 75% of this workforce were employed in firms with less than 50 workers. The benefits of tightly linked small firms which make up the Prato textile district are as follows:

1 **Reduction in the size of firms and sub-contracting** to local firms permitted **specialisation** and a **quick response** to changes in taste and fashion.

2 The industrial district consisted of thousands of firms which could **exchange ideas** and **diffuse innovations** rapidly.

3 The presence of a high level of **professional skills** allowed for development of the product.

4 Small firms encouraged **entrepreneurial activity** and also encouraged **collaboration** and **specialisation** rather than competition.

The success of flexible specialisation and industrial districts in Third Italy has led to calls that a similar approach be used in the South. However, it seems unlikely that with its poor infrastructure, peripherality and low skilled workforce the Mezzogiorno could use a similar path to development.

Figure 7.13 Flexible specialisation: textiles in Prato

Section D The Ruhr

The Ruhr (Figure 7.14) is the classic example of an old industrial heartland in transition. It was once the foremost centre of coal and iron and steel production in Europe. Between 1850 and 1914 the workforce of the Ruhr's collieries increased from about 12,000 to 400,000. However, through deindustrialisation and restructuring both the coal and iron and steel industries have declined. 40% of all iron workers in the old West Germany and 60% of miners were employed in the Ruhr in 1984. Between 1984 and 1994 a fifth of these workers were made redundant (Figure 7.14). The reasons for this decline are common to most of Western Europe: a mixture of competition from abroad, overcapacity and technological change. The results are also similar, involving increased unemployment, massive out-migration and the need for radical restructuring and regional growth policies.

REGIONAL POLICY

The Ruhr is designated an **Objective 2** region (economic decline and high unemployment) and as such has been the recipient of substantial aid from the EU. Between 1989 and 1991 the area received over £270.9 million. However, funds from the EU are increasingly stretched between the peripheral regions. The German government is faced with the huge burden of building a united Germany which means the Ruhr must find different, more locally-based strategies to guarantee jobs for its 5.5 million population. In the late 1980s and early 1990s three initiatives were launched:

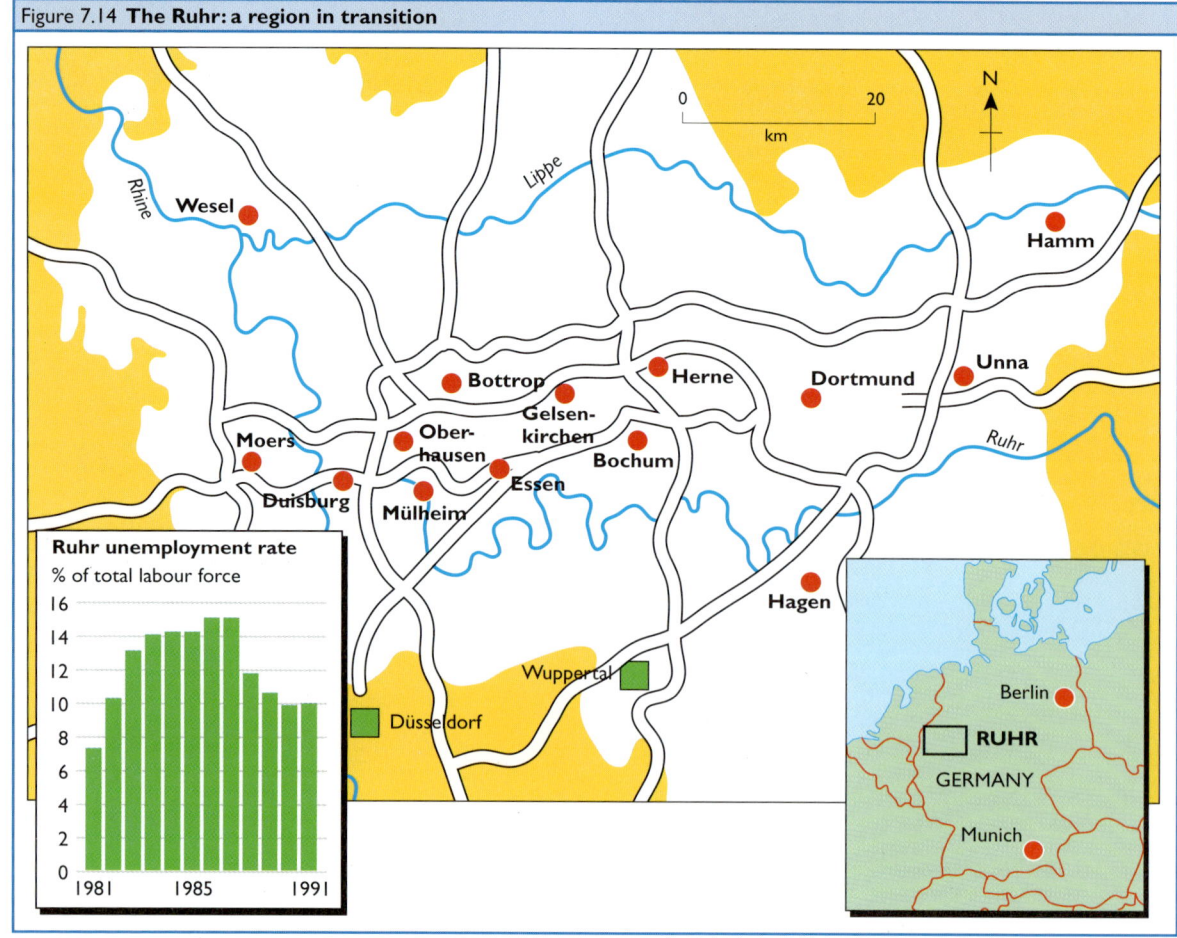

Figure 7.14 The Ruhr: a region in transition

1. 35 businessmen joined forces to found the **Initiativkreis Ruhrgebeit**, each contributing DM 1.5 million to form a lobbying group.

2. The regional government of North Rhine-Westphalia, which governs the Ruhr, launched the **Zukunfts-initiative Montonregionen** (the initiative for the future of the coal and steel regions) to promote innovation, new technologies and new training.

3. The regional government also set up the **Emscher Park Planning Company** in 1988 which is a 10-year programme, covering 17 towns and two million people—the aim is to re-landscape the most depressed part of the region and attract new, high-tech industries to the area.

The result of these and previous policies is that while the coal and iron and steel industries are gradually closed new industries have been developing, particularly in environmental industries and technology transfer centres.

ENVIRONMENTAL INDUSTRIES

In the early 1960s the Social Democratic Party in North Rhine-Westphalia started its regional election campaign using the slogan 'a blue sky over the Ruhr'. It was the first indication that there might be an alternative to the **air pollution**, **water contamination** and **landscape degradation** which seemed the natural cost of economic growth.

The Ruhr's first-hand experience of environmental problems have given rise to a very successful industrial specialisation in **environmental technology** (Figure 7.16):

- More than 150,000 people now work in environment-related industries.
- Between 1990 and 1994, DM 6 million was invested in 4,500 environmental companies in the region.
- It is estimated that the value of the demand for environmental protection in the EU will be about DM 130 billion in 1995.

Ruhrkohle, the coal company, has taken advantage of this opportunity, diversifying into environmental technology. Two of its subsidiaries, Ruhrkohle Umwelt and Steag, are examples of its success. Runkohle Umwelt specialises in waste management and employs 1,500 workers. Steag operates power plants for district heating and related environmental services which employ 4,500 people.

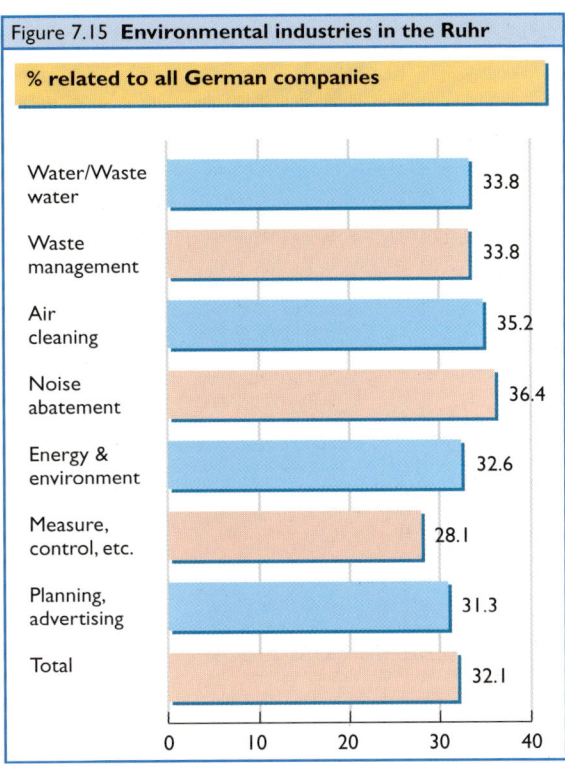

Figure 7.15 Environmental industries in the Ruhr

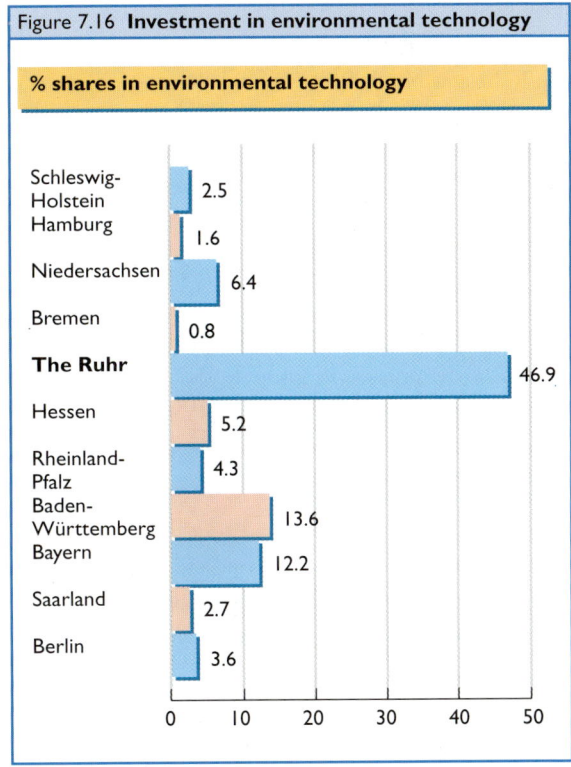

Figure 7.16 Investment in environmental technology

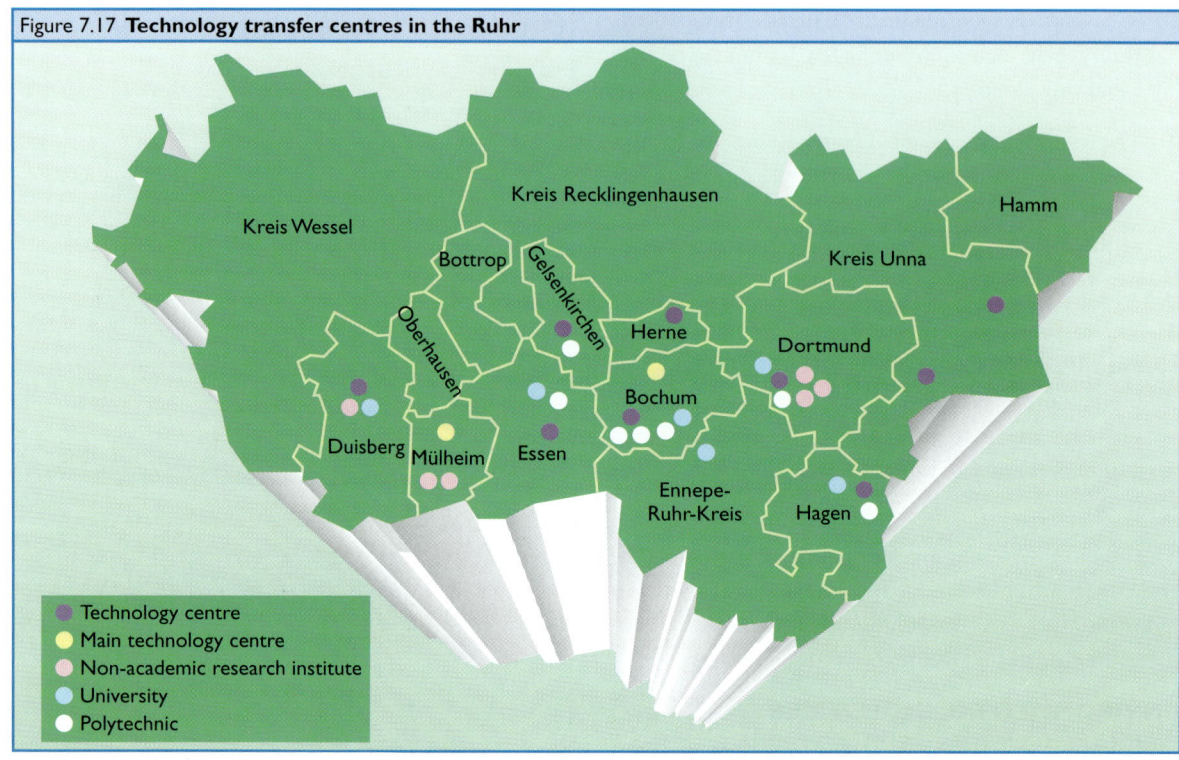

Figure 7.17 Technology transfer centres in the Ruhr

TECHNOLOGY TRANSFER CENTRES

Figure 7.17 shows the network of centres established in the Ruhr and North Rhine-Westphalia to promote new technologies. The **Technology Transfer Centres** (TTCs) are designed to provide marketing advice and also spread new ideas. The TTCs are strongly linked to universities and research institutions of which the area has among the highest concentrations in Germany:

- 39% of technology centres;
- 38% of TTCs;
- 41% of the projects to promote the establishment of new enterprises;
- 30% of the projects on innovative qualification and retraining.

Through TTCs, universities have been used not only to provide alternative employment but also to stimulate the establishment of new industries by breaking through the bias towards coal and iron and steel related growth.

THE FUTURE

The Ruhr has had considerable success in providing alternatives to its deindustrialising heavy industries:

1. Unemployment, which had peaked at 15% in 1987 (from 1% in 1970), declined to 11% in 1993.

2. Net emigration ceased, and a new influx of immigrants, largely from Eastern Europe, reversed the trend.

3. Between 1965 and 1985 total employment in the region fell from 2.37 million to 1.92 million; between 1985 and 1991 a surplus of 237,200 jobs was created.

However, the region still has problems. It has lost close to 100,000 steel jobs since the mid-1960s and EU restructuring could see 50,000 more redundancies. Coal has lost 240,000 jobs over the same period and the reduction in production from 115.4 million tonnes in 1960 to 50 million today would have been even more rapid had it not been for massive state subsidies (the so-called **'coal penny'** imposed on household electricity bills). At the same time it faces competition for new investment and production from the former East Germany.

The Ruhr's regional policy has been successful but the region still faces clear difficulties in converging with more affluent German states.

Figure 7.18 **A development agency publicises the region's new image**

DAS RUHRGEBIET.

The Ruhr. The driving force of Germany.

Research and Transfer

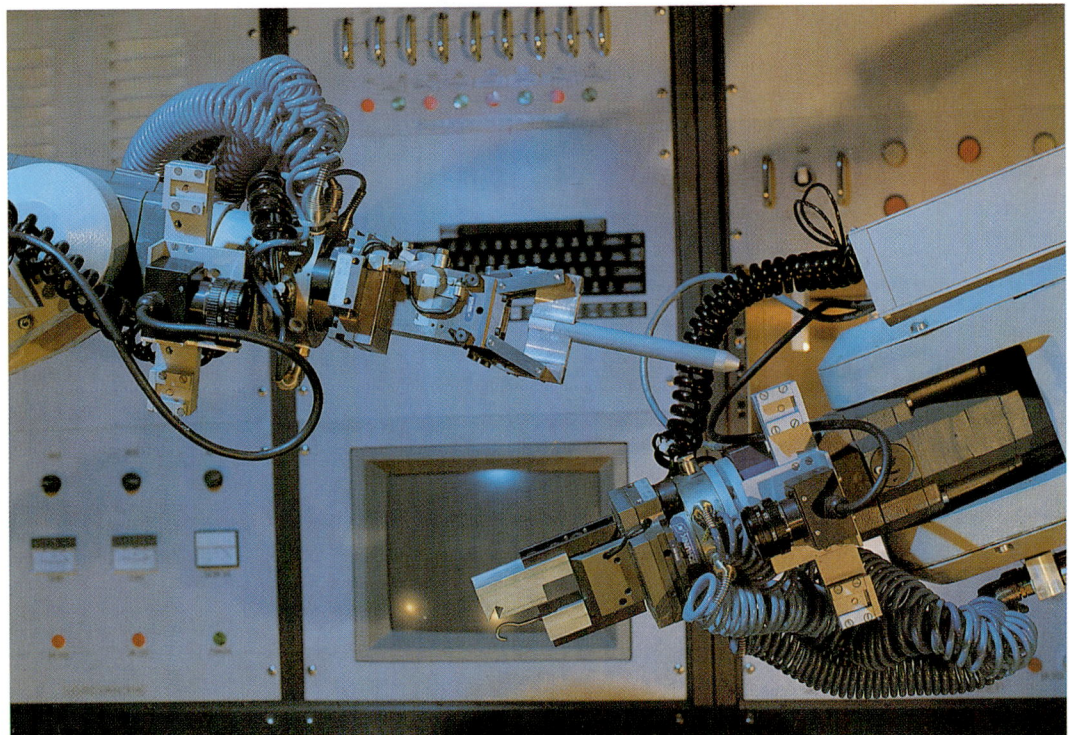

Research and transfer propel economic development. With a diverse economic and industrial structure, the Ruhr is moving into the future - open to cross-border co-operation, project-related work and exchange of experience.

Section E Exercises and recommended reading

EXERCISES

1. Define the following terms which are used in this chapter; *growth pole strategy*; *inward investment*; *an Objective Two region*; *assisted area status* (UK). [8]

2. The table below shows the policies used in the regions covered in this chapter.
   (a) Describe and explain how each policy aims to increase regional prosperity. [10]
   (b) Why is it impossible to use the same regional policy in Europe's regions even when the problems of deindustrialisation are similar? [10]

Figure 7.19 Regional policy

REGION	POLICY
THE UK **West Midlands**	attracting foreign MNCs to problem areas use of regional grants to attract new industries
ITALY **Mezzogiorno** **Third Italy**	growth pole strategy industrial districts and flexible specialisation
THE RUHR	locally-based initiatives based private and regional funding; technology transfer centrs

3. The new map of regional assistance (p. 117) was welcomed by some and condemned by others. Read the two quotes below which were resppnses to the map:

   We have cracked the conventions Whitehall and Tory Party wisdom that ther is no depression or misery in good solid Tory South East England. The area has a completely stagnant economy, a low skill base and a poor infrastructure!'

   Sir Alastair Morton, Chief Executive of Eurotunnel

   'It is no coincidence that the South, South East and South West are the major beneficiaries. It is ... a vain attempt to prop up crumbling Tory support in those areas.'

   Alan Milburn, Labour MP for Darlington (which lost its intermediate status)

   How far do you think the new map was created for political rather than economic reasons? Explain your answer. [13]

4. Choose **one** of the following:
   (a) Regional development policies have been successful in some areas and unsuccessful in others. With reference to specific examples identify areas where these policies have either been successful or unsuccessful and attempt to explain why. [25]
   (b) Identify the obstacles that hinder regional economic development and give examples of how such obstacles have been tackled. [25]

RECOMMENDED READING

An expanded version of the Section on the UK's North-South divide can be found in 'Growing Richer in the Regions' in *Geographical*, January 1994. A good overview of the geography of the West Midlands is Mick Healey's article in *Geography*, 1995, no. 347, part of an interesting series on Regional Development in the UK. Leonardi and Nanetti's (1994) *Regional Development in a Modern European Economy: The Case of Tuscany* is one of the better discussions of the rise of Third Italy. Russel King provides an excellent overview of the Italian regions in 'Italy: from Sick Man to Rich Man of Europe' in *Geography*, 1992. The *Financial Times* runs yearly surveys of Germany, The Ruhr and the North Rhine-Westphalia. Alun Jones (1994) *The New Germany—A Human Geography* is a timely book on recent developments in Germany.

CHAPTER 8
THE PERIPHERY

A **NORTHERN IRELAND AND THE REPUBLIC OF IRELAND**
   Northern Ireland 134
   The Republic of Ireland 137

B **SCOTLAND AND WALES**
   Scotland 140
   Wales 143

C **GREECE, PORTUGAL AND SPAIN**
   Greece 146
   Portugal and regional aid 148
   Spain and EU membership 150

D EXERCISES AND RECOMMENDED READING 151

Section A evaluates the development of Ireland. Northern Ireland is the most peripheral region of the UK. Its outdated economic structure helps explain, to an extent, why it lags behind the rest of the country in terms of living standards. However, this is changing and the 'peace process' holds great potential, on both sides of the border. Ireland continues to make a remarkable transformation from a largely traditional, depressed rural economy to a dynamic, high-technology industrial society. Although some disadvantages remain, notably its lack of resources and its peripherality, it has developed on account of its skilled, adaptable, enthusiastic population and government commitment to advancement.

In **Section B** two peripheral areas of mainland UK are examined. Scotland contains severe regional inequalities: deindustriaslisation in the Central Belt, the impact of oil on the eastern part of the country and the physical and economic isolation of much of the Highlands and Islands. Wales, however, is blighted by an outdated image as a country of smokestack industries and an undeveloped infrastructure. Neither is particularly true, although the country does have poor internal transport links and an ageing population. Wales is seen as having the potential of becoming one of Europe's most prosperous regions largely through inward investment.

Section C investigates conditions in the periphery of mainland Europe. Spain, Portugal and Greece joined the EU in the 1980s and have had a considerable pull on EU resources at the expense of the UK and Ireland.

Section A Northern Ireland and the Republic of Ireland

NORTHERN IRELAND

Northern Ireland is a unique part of the UK. It is characterised by a relatively unspoilt environment, good physical and social infrastructure, excellent communications and a youthful age-structure. However, it is **peripheral** to the UK, has poor transport links with the Republic of Ireland, has high levels of unemployment and there is a marked east-west regional divide. Northern Ireland has been perceived as a **problem region** on account of its specialised manufacturing base, notably shipbuilding and textiles (pp. 72-3), and the recent 'Troubles' of 1969-94. Nevertheless, there are real prospects for growth in the province and this section examines the economic impact of peace, inward investment, aid, and in conjunction with the Republic, cross-border initiatives and tourism.

Economic performance

Ironically, its economic performance in the recent recession of the late 1980s and early 1990s has exceeded the rest of the UK for a number of reasons. The province's **structural weakness** caused the strong performance: the high proportion employed by the government, 40% of the workforce, and the high level of capital availability to firms in Northern Ireland aided the economy. Moreover, the strong performance of Northern Ireland's manufacturing base is explained by its markets: up to two-thirds of products are exported outside of Britain, compared with only about a third for most mainland manufacturers. The underdeveloped nature of the financial service industry and the abundance of housing meant that the boom in house prices that led to the British recession bypassed Northern Ireland. Rents too are lower than in the rest of the UK. Consequently, Northern Ireland's population has a greater disposable income. In addition, in 1994 unemployment fell by 1% and job vacancies increased.

The 'peace process' – economic dividend or deficit?

The potential impact of peace in Northern Ireland is profound although the implications are both positive and negative. On the one hand, businessmen expect up to 30,000 new jobs by the year 2000, 10,000 in **tourism** alone. Currently, this accounts for just 1.5% of Northern Ireland's GDP compared with 7% of Ireland's. US **aid** is set to increase from £12.9 million p.a. to £130 million p.a. and EU aid from £1.076 billion between 1989 and 1993 to £1.23 billion between 1994 and 1999. However, there will be less demand for **security** personnel and linked occupations, increasing unemployment levels in this sector. Up to 20,000 jobs could be lost as a result of peace. Although the cost of 'policing' Northern Ireland, over £1 billion each year, will be reduced, so will the subsidy from Westminster, £3.6 billion.

Inward investment

Since 1986 overseas investment in Northern Ireland has totalled £1.6 billion, involving 265 developments creating 22,702 jobs. These have been attracted by a number of factors: an **educated and skilled labour force** with wage levels 16% lower than those in mainland Britain; **competitive business costs**; IDB **incentives**; a **strategic location**; English, the **universal business language**; its technology **infrastructure**. Thus, a peaceful Northern Ireland holds considerable attractions for foreign investors.

The IDB is able to offer a package of **financial incentives** that include 50% of set up costs and a **comprehensive after-care service**, such as support marketing, **strategic planning** and **targeting** of potential suppliers. Northern Ireland can offer 50% cash grants for buildings and grants for up to 100% of rental costs for up to five years, more generous than either Scotland or Wales. Land and property is considerably cheaper than in mainland Britain. Industrial land costs £40,000 an acre compared with £185,000 in the English Midlands and £400,000 in London.

New investment plans announced at the **Northern Ireland Investment Forum** in December 1994 by John Major, the prime minister, confirmed five new investments totalling £74 million:

1 Du Pont (USA) is developing prototype technologies.
2 Ford is to invest £15 million in the production of new engine components.
3 Nacco Materials Handling of the USA will be developing local materials for forklift trucks.
4 British Telecommunications (BT) is investing in a new office block in Belfast and expanding its customer services centre in Enniskillen.
5 Fujistu is to build a £3.5 million electronics factory in west Belfast.

Figure 8.1 **The Emerald Corridor**

Cross-border initiatives: the Emerald Corridor

The low levels of trade and cooperation between Northern Ireland and the Republic of Ireland is due to historical and infrastructural reasons. The **Single Market** in 1992 caused many businesses on both side of the border to improve their trading relations. Although the Single Market removed delays at customs, the state of the road and rail network remains poor. However, these links are expected to improve substantially between 1995 and 1999 as they are a priority in the Republic's IR£7 billion National Development Plan.

One area which holds out great potential for regional economic development is the **Belfast-Dublin corridor**, the **'Emerald Corridor'** (Figure 8.1). Here, 50% of the island's population is concentrated on 20% of the land. If peace, or even stability, were to come to Northern Ireland, up to 75,000 jobs could be created in this region. Such a corridor would give Northern Ireland companies better access to the island's largest concentration of consumers. Already several firms have made use of this **core area**, with its improved communications and access to the main Irish and Northern Ireland markets. It also access to mainland Britain and the enhanced European Union market of over 350 million people.

The Emerald Corridor has many other **attractions**. For example, in the UK, Larne is second only to Dover in terms of the amount of ro-ro (roll on-roll off) commercial traffic handled. Larne's importance in terms of regional growth cannot be disputed. It handles over 500 container vehicles daily, and up to 20% of traffic in the port originates from or is destined for the Republic. It has good road connections and access to the main markets north and south of the border.

Tourism

In December 1994, the British and Irish governments announced plans to remove partition from the tourist map by marketing Ireland as a whole. The governments hoped to attract 92,000 more visitors in 1995 with a £6.8 million tourism **incentive**. A further £4 million will be provided by international aid. In 1993 1.25 million people visited Northern Ireland and 10,000 were employed directly in the industry. By contrast, in Ireland there were about 3.5 million visitors annually (a tourist for every citizen) and over 90,000 were employed, making it the Republic's third most important industry. Hence there is considerable scope for promoting Northern Ireland's tourist industry and up to 30,000 new jobs could be created. However, many of the jobs in tourism are seasonal, low paid and relatively uncertain (**pp. 96-7**).

Some **developments** in tourism are quite substantial and reflect the growing confidence of investors in the province. **Hilton International** plan to build a £17 million luxury hotel along the Lagan River in Belfast. The Belfast Hilton will be part of a £130 million investment scheme in the Laganbank development site just 500 metres from City Hall and the main shopping centre. This scheme will include a concert hall and conference centre, new office space, restaurants and public houses.

Figure 8.3 **The Giant's Causeway, Antrim**

Figure 8.2 **Carrick-a-Rede, Antrim**

Aid

Since the Downing Street Declaration and the paramilitaries' ceasefire there has been a substantial increase in the amount of aid allocated to Northern Ireland. However, there has been considerable aid to the province for many years. The economy has been protected from the worst of the recession by an annual £3-4 billion **subsidy** from the British Exchequer, and the saving in security spending over 1995-7 is to be used for local health and education budgets.

The EU has also indicated that there will be considerable funding available for Northern Ireland. It has adopted a **development programme** of ECU 2.4 billion for 1994-9 and proposed an ECU 300 million support programme for peace and reconciliation and is also proposing to increase the contribution to the **International Fund for Ireland** by a third to ECU 60 million. The aid will be divided into two categories: **promotion of economic development and competitiveness**, and **investment in human resources**. This is designed to promote urban regeneration and better relations between the communities in Northern Ireland. In addition a further ECU 98 million is available from the EU's INTERREG programme for cross-border cooperation between Northern Ireland and the Republic between 1994 and 1999.

THE REPUBLIC OF IRELAND

By contrast to Northern Ireland, Ireland's development since 1969 has been very different. It has been transformed from a primarily agricultural and agricultural-processing economy into a vibrant high-tech economy. Unlike Northern Ireland it lacked a manufacturing base but shared a poor location relative to Europe. Indeed Ireland is the only member of the EU without a land-link to other member states; it has few mineral resources of any value and it has a small home market. Ireland has depended upon **government initiatives** in order to modernise, and, as it has developed, there has been a change in emphasis from **import substitution industries** to **export orientated industries**. Regional policy has been limited to times of economic boom and has generally been submerged under strong centralising forces and national interest. It is now one of the most buoyant economies in Europe with a trade surplus equivalent to 13% of GNP (1993), above EU average GNP growth rates (4-5%) and low inflation rates (≤2%). Although Ireland may remain one of the poorer members of the EU its relative prosperity is increasing: average income has risen from 60% of the EU average in 1985 to 74% in 1994.

Manufacturing change

The government's policy of attracting industry with a generous package has brought about a remarkable upturn in Irish economic fortune. Manufacturing replaced agriculture as the dominant element of the economy. **Globalisation** by multinational corporations established many branch plants in Ireland, thereby attracting firms eager to infiltrate the European market. The **division of labour** into production and research/decision-making enabled the government to steer many plants into the rural areas and small towns. This was more noticeable during the growth period of the 1970s. During the boom years of the 1970s Ireland became a country of immigration rather than emigration, and population growth was widespread throughout the country. Consequently, regional inequalities lessened, only to re-emerge in the 1980s due to the recession.

Since 1981 manufacturing decline has affected all parts of Ireland. The reasons for decline varied spatially. In the core, it was the decline of the traditional indigenous industries, textiles, clothing and footwear, owing to a failure to restructure and overseas competition. The poor performance of some MNCs globally caused them to **rationalise** and close branches, frequently in the **periphery**. Consequently, the government's main concern was with national growth rather than with regional inequalities, hence its emphasis on sectoral interests.

Trade

The effect on Ireland of the enlarged EU market is of vital importance. Ireland depends heavily on external trade, notably with the UK, EU and USA. Two-thirds of its exports are manufactured goods, mostly from MNCs. By contrast, indigenous firms are generally small-scale and orientated to the Irish and British markets. Prior to joining the EU in 1973 Ireland was heavily dependent on the UK as its main market, accounting for 65% of total exports in 1970. However, by 1993, 76% of exports went to non-UK destinations. Moreover, there have been substantial changes in the components of these exports. In 1980 food, live animals and drink accounted for 36% of exports, the largest share, whereas by 1993 this had declined to 21% of exports and was second to electronics, 27%.

Figure 8.4 **A view of Dublin's financial centre**

The Industrial Development Authority (IDA)

The IDA has taken a high profile in attracting inward investment to Ireland. Although critics bemoan the lack of support for indigenous firms, it must be recalled that Ireland is not a rich country and therefore must seek international capital, in the form of MNCs. Over 1,000 overseas enterprises operate in Ireland, 370 from the USA, employing over 90,000 people.

Ireland has a number of **advantages** for the investor. It has advantages in its **highly educated**, **skilled**, **multilingual**, **youthful workforce**, **healthy economic indicators** and its ability to continuously attract foreign investment. Between 1970 and 1992 Ireland's population (3.5 million) increased by 20%, over twice the EU average: 45% are under the age of 25 years, compared with the EU average of 33%, and 27.5% of the population are in full-time education. **Competitive labour costs** also make the country very attractive for investors.

The success of the new investments also depends on the quality of the **telecommunications infrastructure**, and during the 1980s Ireland invested $3.5 billion to create one of the world's most advanced telecommunications networks. The **International Direct Dialling System** offers access to 90% of the world's telephone users and satellite, copper- and fibre-optic systems provide global links through a full range of telephony, telex, digital data lines, packet switched data and voice private wires.

Peripherality

Nevertheless, there are problems which continue to affect the Republic. **First**, Ireland is now the only EU member without a **land-link** to mainland Europe. **Second**, the **internal market** of 3.5 million is not enough to sustain industrial development and dependence on Europe, especially the UK, is a fact of Irish economic life. **Third**, Ireland's **high birth rate** means there is a huge dependency on wage earners. It is the

Figure 8.5 **Killarney National Park**

Over 1 million visitors travel to Kerry annually, bringing £160 million into the country. Killarney is one of the major tourist attractions in Ireland. To meet the conflicting needs of tourism and conservation in Killarney National Park, a carefully managed land-use plan has been developed.

highest in Europe, between 200% and 300% compared with 130% in Denmark. Consequently, the need to generate employment in Ireland is imperative: the rate of increase in employment opportunities needed to keep pace with population growth is fifteen times greater than in the rest of Europe. **Fourth**, Ireland has had a high rate of **unemployment** for over a decade. Despite the growth of so many firms in Ireland, unemployment remains a great problem. Apart from Spain, Ireland has the highest rate of unemployment in the EU and over 40% of Ireland's unemployed are long-term (over one year). Unemployment increased dramatically during the recession of the 1980s and early 1990s and was associated with increased plant closure and the failure of new employment opportunities to fully soak up the unemployed. In part the problem was offset by the high rate of emigration from the Republic to Britain and the rest of Europe. Unemployment thus became a problem not only of the peripheral north and west but also locally in the inner cities. **Fifth**, the west of Ireland has suffered disproportionately from **reform** of the EU's Common Agricultural Policy. This has forced many in the traditional dairy and beef cattle industries to look for alternatives in new industries. **Sixth**, Ireland amassed enormous **debts** in the 1970s and is still trying to repay them. It is one of the world's major debtor countries and has one of the highest debt: GNP ratios (72% in 1985, 90.9% of GDP in 1993). **Seventh**, it remains **dependent on external capital** and is extremely sensitive to rationalisation and cut-backs of the MNCs. Dependency on multinationals and a branch plant economy is beset with problems. There may be little multiplier effect and large-scale dependence can prove disastrous when an employer decides to pull out, as in the case of Digital closing down its Galway operations in 1993. An estimated 40,000 people (workers and their dependents) were effected by the plan to concentrate production in Scotland.

Finally, **regional inequalities** show little sign of declining: indeed, the lack of coordinated planning and the sectoral approach adopted by the government have only served to enhance variations in the socio-economic conditions throughout Ireland. Ireland exhibits clearly the functioning of the core-periphery model. There has been limited attempts to reduce the disparities between regions, and government policy has generally favoured national interests. However, the contrasts are not just south and east versus north and west—there is an urban-rural dimension and within the urban areas an inner city-suburban dimension. Within the rural areas there is socio-economic polarisation between the 'traditional' farmers and the emerging middle-class commuters. In effect, Ireland illustrates most of the spatial variations to be found within any developed country.

Ireland's position in Europe is not as disadvantaged as it is often made out to be. On the other hand, there is little room for complacency and only a vigorous policy of attracting industry and educating its workforce has enabled Ireland to arrive at its current position. Its comparative advantages are recognised, and vigorously advertised by the IDA, and the economic and political stability of the country provides a favourable climate for inward investment by companies eager to reach the European market.

INSET 8.1 INVESTMENT ADVANTAGES

Ireland has a number of advantages for the investor:

1. *A unique and beneficial tax system for industry: only 10% corporation tax to the year 2010.*
2. *A plentiful supply of well-educated, multilingual young graduates and skilled workers: 50% of the population is under the age of 28—in an ageing Europe faced with a shortage of skilled labour this gives Ireland a distinct advantage.*
3. *A generous range of government capital, employment, R&D grants and other incentives: capital grants for plant and machinery and a 100% grant on training.*
4. *Low inflation and labour costs.*
5. *Assistance from IDA—the government industrial development agency.*
6. *A state-of-the-art digital telecommunications system.*
7. *Duty free access to the EU market of 370 million consumers.*
8. *A return on investment at 25%, almost four times the EU average.*
9. *A unique quality of life with superb sporting and leisure facilities.*

Section B Scotland and Wales

SCOTLAND

Scotland accounts for approximately one-third of the land area of the UK and about 9% of the UK population. The country divides into **three distinct areas**: the Southern Uplands, the Central Belt where the bulk of Scotland's people and economic activity lie, and the Highlands and Islands, which includes the Eastern Belt, an area largely affected by North Sea oil.

There are wide variations within Scotland in terms of **unemployment**. It is lowest around Aberdeen and the Shetlands, both associated with North Sea oil, and highest around areas such as Greenock, Lanarkshire and Alloa, where Scotland's traditional industries dominated, as well as remote islands such as the Western Isles. The cities of Glasgow, Dundee and Edinburgh and the depressed urban areas of Lanarkshire account for almost 60% of Scotland's unemployed and male unemployment rates of 25% are not uncommon.

The economy

Agriculture, forestry and fishing contribute about 3% to Scottish GDP and are of greater importance outside of the Central Belt. Almost 75% by volume of the fish landed in the UK is at Scottish ports. Forestry provides over 15,000 jobs in rural areas. **Manufacturing** contributed a third of Scottish GDP in 1966 but less than one-fifth in 1989, while the share of services rose from a half to two-thirds. The most important production industries include the extraction of oil and related oils and chemicals processing industries; electronics and high-technology engineering industries, which have been the fastest growing manufacturing industries since 1981; whisky, which continues to be a major manufacturing earner; and high quality textiles, which is a key manufacturing industry in many rural areas. During the 1980s **electronics** output quadrupled and employment increased by 12% at a time when manufacturing employment as a whole in Scotland fell by 30%. Electronics and instrument firms are centred on Silicon Glen, the area between Glasgow and Edinburgh.

The Central Belt

The economic transformation of the past forty years, especially in and around Glasgow, has led to periods of high **unemployment** and high **out-migration**.

Strathclyde is by far the most populous region and contains the main **concentration of heavy industries**. From the early 1960s, it experienced **deindustrialisation**, more rapid and severe than elsewhere in Scotland. The **Ravenscraig steel plant** (pp. 66-7) closed in 1992 and almost all of Scotland's heavy industries are only memories. Nevertheless, Strathclyde is set to be the location of a major British Rail Eurofreight terminal for Channel Tunnel traffic, to be situated in Mossend, Lanarkshire. Scotland, having done relatively well out of defence expenditure over the decades of the Cold War, is now suffering from the **defence cuts**. The Rosyth Naval Dockyard in Fife has seen its workforce drop from 6,000 in 1988 to 3,600 in 1994. In 1993 the government awarded the contract for servicing Trident submarines to Devonport instead of Rosyth, and its workforce is set to drop to 2,200 by 2005.

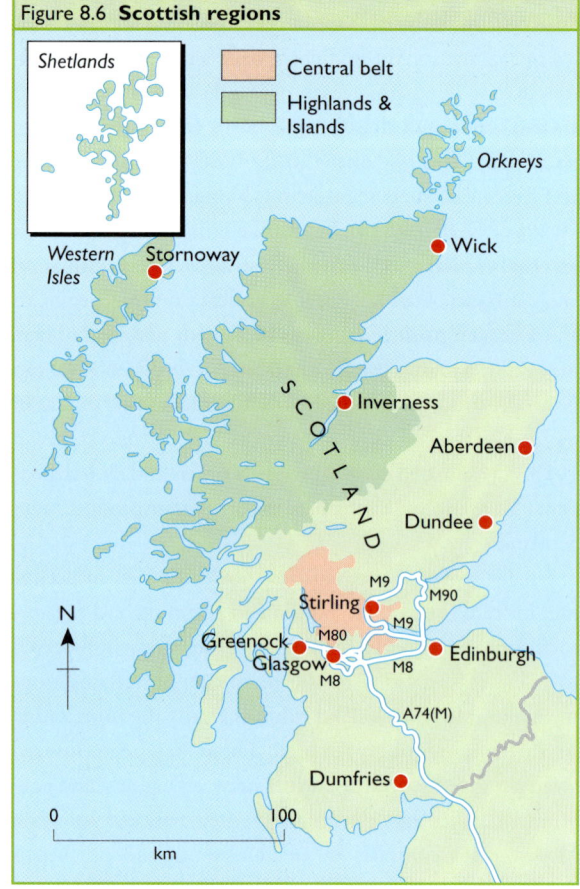

Figure 8.6 **Scottish regions**

Figure 8.7 High-tech industry in Scotland: IBM

Inward investment now accounts for the predominant part of Scotland's electronics industry and a very important part of the North Sea oil related activity. Scotland's dependence on inward investment is starkly underlined by the fact that the electronics industry, which consists largely of foreign-owned branch plants, lacks the ability to respond spontaneously to market opportunities. It is a sign of just how much the Scottish economy has changed in the past few years that the sector spearheading the economy is the largely foreign-owned computer industry. Financial services are also expanding, especially in Edinburgh.

Grampian: the Eastern Belt

Until the discovery of North Sea oil, farming and fishing formed the basis of the region's relatively diversified economy which also included food processing, distilling, textiles, paper-making and shipbuilding. Although many of these have declined in recent decades the **oil industry** now plays a central role in the region's economy, creating a peak of about 70,000 jobs in the early 1990s or one-third of the region's employment.

Aberdeen dominates the region, with almost half of the population and two-thirds of the workforce. In other towns, the industrial structure is very different, notably due to the lack of **oil-related employment**. Oil revenue has enabled **diversification** of employment and an increase in the social and physical infrastructure of the area. Aberdeen is very much a **regional growth pole** and has the world's busiest heliport at Dyce airport. Intense lobbying of the DTI helped relocate part of the petroleum exploration directorate to Aberdeen and there is a continuing movement of oil-related companies to the north east simply because it is still cheaper than South East England and more cost-effective to be near the activity. On the other hand, oil-related activities may have led to the **decline** of some traditional activities such as **textiles**, increased the cost of living in the region and created serious environmental damage.

Oil has benefited the British economy in a major way: it is a reliable source of energy, the energy base has been diversified, trade balance has improved and it has created numerous jobs and a growth zone in eastern Scotland. However, how much of the financial returns should remain in Scotland is a contentious issue. Only about one-third of the 27,500 offshore workforce live in the area: most live in central Scotland, North East England and further afield, causing much of the earnings from oil to be spent outside the region.

Highlands and Islands

The more **peripheral areas**, including the Borders and parts of the Highlands and Islands, depend heavily on tourism, agriculture, fishing and textiles although the development of North Sea oil has certainly benefited some parts of the region at the expense of others. The area is a large sparsely populated region, with less than 9 people/km², i.e. 15% of the UK land area but less than 1% of the population. Although it is an area of striking beauty and unspoilt scenery, and is therefore a major attraction for tourists, the geography of the region presents special problems for transport and communications. Its development is hampered by its **peripherality**, its small and **scattered population**, its **small market** and **labour force**, and by the geographic and climatic characteristics of the area.

A high proportion of the population is engaged in **primary industries** and in **tourism**, and almost all of the area is designated by the EU as a Less Favoured Area (LFA). The Highlands and Islands is one of the EU's poorest regions and qualifies for maximum economic assistance, as much as £260 million between 1995 and 2000. The region is very remote and arable land accounts for just 2% of the area.

Tourism accounts for about 20% of the region's GDP and an increasing part of the workforce. However, it is vulnerable to short-term climatic and economic changes. Salmon farming employs some 6,000 people. Fishing and fish-farming are particularly important, with the region accounting for over half of the fish landed in Scotland. On the Western Isles, for example, most of the working population are engaged in fishing, farming and textile manufacturing. Incomes are low and unemployment is high; the remoteness of the islands and the dependence on ferry services add to the economic hardships of the area.

Industrial development in the Highlands and Islands has included both large-scale and small-scale manufacturing. A combination of North Sea oil, HEP, timber and water resources have attracted many firms to Inverness and the Moray Firth. Construction work related to North Sea oil has been concentrated in the area. Europe's largest oil reception, treatment and trans-shipment facility is situated in the Shetlands. Exploration, development and production of North Sea oil and gas have made an important contribution to some local economies, especially the Orkneys, Shetlands and the inner Moray Firth. However, this activity has passed its peak and is subject to variations in activity. However, not all large-scale manufacturing has been successful: the 6,000 workers brought into the Shetlands (population 18,000) has created problems of integration socially, economically and culturally. The over-dependency upon oil became evident with the rapid rundown of the two large oil rig construction yards which had brought considerable prosperity for most of the 1970s and 1980s to communities around the Moray and Cromarty Firths.

Figure 8.8 **The effect of oil on the environment: the Braer disaster, January 1993**

WALES

Although 80% of the country is **rural** only 5% of the workforce, 65,000, are employed in agriculture. Welsh farming is typical of **peripheral areas**: it is extensive and only 3% of holdings are used for cropping. The **heavy industries** of coal and iron and steel formed the backbone of the Welsh economy for over 200 years. However, this has changed. In 1920 over 250,000 people were employed in the coal industry, now it is less than 2,000. By contrast, although employment in **iron and steel** has also contracted severely, from 85,000 in its heyday, it still accounts for about 14,000 jobs and 30% of UK production. More recently, the **closure of defence industries** and air bases in west Wales has had a detrimental impact on the local economy. However, growth has occurred in the energy industries such as oil refining and electricity generation.

Parts of Wales experience difficult demographic conditions:

1. A higher proportion of **elderly people** than anywhere else in the UK.
2. Considerable **out-migration** from industrial areas.
3. A **shortage of skilled labour**.

High unemployment and **low incomes** characterises many of the **former coalfields** as well as the extreme north west and south west. In Wales as a whole gross earnings are only 89% of that of the UK as a whole. **Average GDP** per capita ranges from 30% below the UK average in Mid-Glamorgan, the area with the highest **unemployment**, to 3% above in South Glamorgan. The GDP/head is lower than anywhere except Northern Ireland, and wages are among the lowest in Britain. The highest unemployment rates are found in the former coal mining valleys of South Wales and in the far west of both North and South Wales, where tourism is seasonally important. In some areas changing global relations has had an adverse effect on employment: hundreds of jobs were lost at the Atomic Weapons Establishment in Cardiff and the Llanishen warhead factory.

Industrial change

Wales has been transformed from an economy dependent on coal and iron and steel to one in which new industries, especially inward investors, have enabled the manufacturing sector to retain a quarter share of the GDP, higher than

Figure 8.9 **Wales**

the UK average. Wales was the only UK region to increase manufacturing's share of GDP between 1980 and 1992. By 1995 there were over 350 overseas-owned manufacturing plants in Wales, and employment by these has risen by 50% since 1985.

Wales has survived better than some other regions because the industrial base has been rebuilt over the last ten years. Wales is seen as being among the first of the UK regions to come out of recession although growth is from a relatively small base. Manufacturing accounts for 27.3% of Welsh GDP whereas the UK figure is only 22.4%. Over 25,000 work in **electronics** and over 70,000 in the **financial services** sector. Although coal is no longer important iron and steel is: Port Talbot and Llanwern are among the most efficient and most profitable steel plants in Europe. Aerospace has also become a sizeable industry employing over 11,000, spurred recently by the British Airways £75 million investment in its maintenance base at Cardiff-Wales airport and £23 million component repair business at Llantrisant. BA has built the largest hangar in Europe to service its Boeing 747 Jumbo fleet.

Despite recent changes in structure, industry remains concentrated in the south and north east. Recent improvements in **communications**, a ready supply of **labour** and an attractive **environment** have attracted many inward investments, especially electronics. The main transport links run parallel to the North and South Wales coastlines and both have been improved recently. The second Severn crossing and the £650 million upgrading of the A55 in North Wales are striking examples of the infrastructural developments benefiting Wales. However, North-South routes are difficult and communications in rural areas are poor.

Inward investment

There are more than 350 foreign-owned companies operating in Wales. These have had a varied impact on the local and national economy. For example, a Japanese-German joint venture between **Nippon Electric Glass** and **Schott Glaswerke** will make glass for cathode ray tubes for televisions. This represents an investment of nearly £200 million at Cardiff creating up to 750 jobs. Site preparation began in 1994 and production in 1995. The first customer was the **Sony** television plant at Bridgend, which previously imported the components. The factory is the most important won by the **Cardiff Bay Development Corporation**, which is responsible for regenerating 2,700 acres of old dockland (**Figure 8.11**). NEG and Schott received regional assistance from the Welsh Office.

Fears that the world recession would make Wales vulnerable to cutback of branch-plants appear overstated although there has been retrenchment. Non-Welsh businesses are continuing to invest in their businesses in Wales. Sony opened a frame-making operation at its Bridgend television plant increasing its Welsh workforce to about 2,700. **Nimbus** manufacturing at Cwmbran, Gwent, made the first compact disc in Britain in 1984 and now produces over 35 million discs annually. It employs 300 people and is investing £2.5 million in its plant. However, one frequent criticism of many inward investments is they only provide low-paid assembly-type work in **'screwdriver' industries**.

Tech-Board, a £40 million start-up company, announced plans to site a hardboard mill, employing 200 people in Ebbw Vale in Gwent. The region's strong manufacturing heritage and the availability of a quality, highly motivated workforce played a major part in the decision. The excellent communications with the rest of the UK and continental Europe, via the Heads of the Valley road and the M4, were also important. Other key factors were the support of the Welsh Office, local authorities and the Welsh Development Agency. Similarly, **Ascom**, a Swiss producer of telecoms, favoured Wales on account of its **flexible labour force**, **good labour rates** and a **workforce with an ability to learn new skills**. Employment costs in

Figure 8.10 Tourist potential: Begwyns, near Glasbury

The Welsh tourist industry is changing and expanding. Traditional seaside holidays have declined and the Welsh tourist board are promoting cultural breaks, hiking trails and wilderness areas.

Figure 8.11 **Redevelopment of Cardiff Bay**

Wales are very **competitive** against those elsewhere in Europe. By contrast, in 1991 **Bosch** invested £100 million in an advanced factory outside Cardiff. Bosch chose Wales on account of its proximity to the M4, the ports of Cardiff and Southampton, and because they wanted to be close to Nissan, Honda and Toyota which had set up in the UK. In general only a small number of foreign firms conduct R&D in Wales, such as **Sharp** of Japan (word processing technology and microwave technology) and Valeo of France (automotive components). Although Wales has a strong element of R&D in medical, biotechnology, electronic and automotive companies, it lacks pharmaceutical and government research facilities which have contributed so significantly to growth in the South East of England.

Peripherality

However, despite these impressive achievements, outside Wales there remains a **perception problem**. It has the lowest regional income of any region and it has not advanced because its development has been based on quantity rather than quality of jobs. It needs to develop a highly educated workforce and it needs to foster small and medium sized enterprises. Moreover, there is general concern that the ownership of many Welsh companies is moving beyond its frontiers. Indeed, relatively few people who run small businesses or tourist facilities are Welsh.

Tourism

Until recently, tourism in Wales was **low-key, fragmented** and **under-capitalised**, consisting mostly of the traditional family summer holidays. Two factors have led to dramatic changes: first, the **decline of the traditional holiday** at the seaside resorts and, secondly, **improved tourist infrastructure**. Wales has had to offer fresh attractions. There are forest walks, heritage parks and industrial and craft museums.

Up to 9 million visitors per year visit Wales and tourism now employs 95,000 or 9% of the workforce, earning £1.3 billion for the Welsh economy. **'Tourism 2000'** plans to create a further 10,000 jobs in tourism and raise its contribution to the economy to over £2 billion per year. There has been a steady growth in **off-season short holidays** and breaks but Wales has fared poorly in attracting foreign visitors, with under 4% of the UK total compared with 9.5% for Scotland. Over £136 million has been spent on greening the countryside and developing tourist attractions, and hills and parkland now covers most of the old coal tip scars. Development programmes are planned for coastal resort regeneration in Tenby, Porthcawl and Llandudno; in the historic town of Caernarfon; and for country holidays, golf, cycling and walking holidays (**Figure 8.10**).

Section C Greece, Portugal and Spain

In addition to Ireland there are three other genuinely peripheral states in the EU: Spain, Portugal and Greece. Each country is locationally disadvantaged within the EU:

1. **Spain** is part of zone of Mediterranean economies where economic dependence on agriculture is large, unemployment is high and incomes per head are less than 75% of the EU average (Catalonia, the French Midi and Northern Italy are exceptions).

2. **Portugal** is part of a maritime Atlantic arc of states and regions which stretches from the Shetlands to Gibraltar with concentration of economic activity in coastal areas and levels of GDP per head of around half of the EU average.

3. **Greece** has developed outside the core areas of North West Europe and is poorly linked to the EU: unlike Spain and Portugal it suffered a fall in prosperity during the 1980s.

These three countries are politically, economically and culturally strikingly different. However, they face similar problems:

1. The Single Market means their agricultural sectors are subject to competition from the better organised and more efficient farmers in the core.

2. They must meet the convergence criteria on inflation and interest rates set by the EU for economic and monetary union (EMU) by the end of the decade.

3. They have considerable regional imbalances within their countries which have been increased by the allocation of Structural Fund money which is often used to develop the most favoured regions.

For geographers, each of these countries provide excellent examples of regional policy and industrial processes which can be contrasted with those in the countries of the core.

GREECE
Greece and globalisation

Since its entry into the EU in 1981 Greece has faced increasing pressure to restructure its industries in a global context. However, it does have some relative advantages: it has **low labour costs** and plentiful **availability of some resources** for food processing, textiles and energy. However, with its entrance into the world economy Greece has seen its comparative advantage eroded away. A major problem stems from the countries of **South East Asia** which have emerged as economic players with an even lower cost base than Greece. But Greece also has structural disadvantages. It is **peripheral** within the EU with no common border with another member state. Also Greek manufacturing has a **low technological capacity, low productivity** and **low competitiveness** by international standards.

Traditionally Greece has built its international specialisation and competitive advantage around three poles: **agriculture, light manufacturing and services**. In the Greek context all these activities are largely low-skill and low-technology. The 1980s saw these competitive poles begin to fail as imports began to penetrate into the economy as result of trade liberalisation. The impact of globalisation was not only to increase competitive pressures on Greece but also to force firms to reorganise and restructure in terms of R&D, industrial relations, training and quality of product. As a result there has been an increasing move towards regional specialisation. The case of Northern Greece is a good example.

Specialisation in Northern Greece

Figure 8.12 shows that Northern Greece can be split into three sub-regions dominated by the core region, Greater Thessalonika. These areas have clear industrial specialisations:

1. **The Core.** Greater Thessalonika developed as an industrial region in the 1960s with a growth in technologically advanced industries such as chemical products and oil refining. It has the most diverse industrial structure but its proximity to the conurbation of Thessaloniki has allowed it to specialise in servicing and assembling and firms producing electrical goods.

2. **Sub-region A.** The coastal areas to the east of Thessaloniki have seen a progressive concentration in the garment industry with an increase in firm size but fewer manufacturing units.

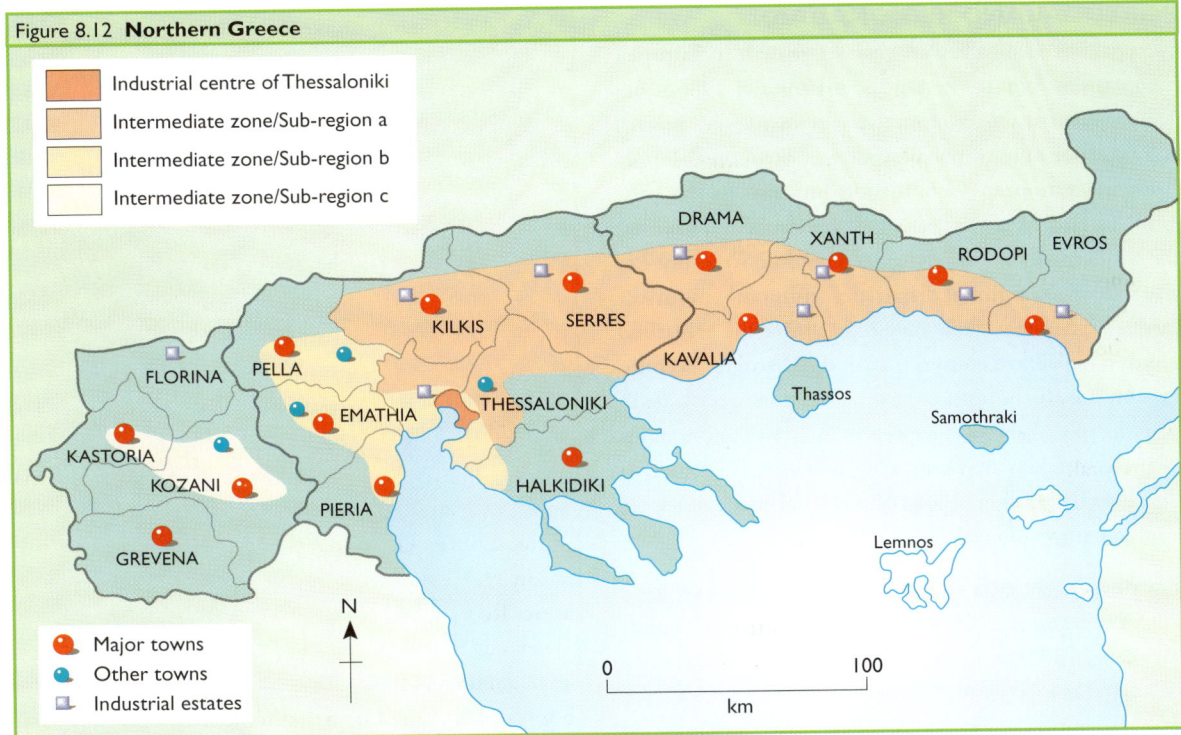

Figure 8.12 **Northern Greece**

3 **Sub-region B.** The areas west and north west form a sub-region which specialises in the food processing industry and textiles. The textile industry is capital intensive with technological innovations introduced in the 1960s and 1970s.

4 **Sub-region C.** The areas of Kastoria and Kozoni specialise in fur and leather products. Kozani has seen an expansion in the energy sector and coal industry largely through government investment.

The location of firms in these regions is due to a number of reasons. A survey has been conducted asking 109 firms in Northern Greece about their reasons for locating. The three areas, **Greater Thessaloniki** (GT), the rest of the **prefecture of Thessaloniki** (REPT, i.e. the rural part of the prefecture), and all other prefectures give some interesting results.

Accessibility to the local consumers market (33%), **proximity to an industrial estate** (21%), **ownership of land** (19%) and **infrastructure** (18%) were important in all the areas. However, the labour market was much more important in GT than in the rest of Northern Greece due to the number of labour-intensive industries. In REPT the existence of regional incentives was much more important than in GT.

The future

The specialisation of manufacturing in Northern Greece and the associated locational influences are being increasingly challenged by the global nature of modern industry. The result is a series of opportunities and threats to the region. Two scenarios have been suggested by the geographers Petmesidou and Tsoulouvis (1993):

1 A **pessimistic scenario** which sees pressure from the European Union to control inflation and competition from NICs and other Balkan States leading to Greek firms closing down, leading to growing unemployment and social conflicts; the result will be considerable regional disparities in both wealth and employment.

2 A more **optimistic view** whereby Greece overcomes its present crisis because its participation in the EU will give it a better position as to the international division of labour than the countries of South East Asia and the Balkans; Structural Fund money will be used to upgrade its traditional industries and attract high-tech activities.

PORTUGAL AND REGIONAL AID

Portugal has been called **'an oasis of growth in Europe'**. Since its accession to the European Union in 1986 it has achieved remarkable economic development. This growth is linked not only to the allocation of **Structural Funds** from the European Commission but also to the pro-growth strategies of the Portuguese government. However, the modernisation of Portugal's economic structure has also influenced **regional disparities** within the country. It can be argued that while investment from the Structural Funds have led to a convergence of the Portuguese economy towards the European norm, it has done very little to help the more backward regions in the north and interior of the country. If anything, after more than a decade of European development aid, the regional inequalities are even more pronounced.

Economic growth

Portugal received 17.5% of the EUs Structural Funds between 1986 and 1994. The amount was doubled to ECU 18 billion between 1994 and 1999. There has been a dramatic impact on the country. Every year since it joined the EU Portugal has achieved its target of growing by 1% more than the EU average. Unemployment has halved and GDP more than doubled since its accession. However, one could also argue that the Structural Funds have created a dependency culture in two ways:

1. The government now clearly expect any large infrastructural projects to be at least partly financed by the EU.

2. In the sectors where Structural Funds are available, business is becoming increasingly dependent an EU funds.

Industrial restructuring

Portugal must **restructure** its industrial base by letting uncompetitive companies die while it saves EU support for industries which will be internationally successful. Restructuring is expected to lead to the closure of a third of the country's plants in textiles, footwear and clothing with the loss of 60,000 jobs by the end of the decade. The hardest hit area is likely to be the **Northern Vale do Ave** textile region, around Oporto, where it is estimated that 120 firms, with more than 30,000 workers, face closure.

At the same time, it is the core areas which are deriving the most benefits from industrial expansion, particularly from foreign investment. **Setubal**, south of Lisbon, has

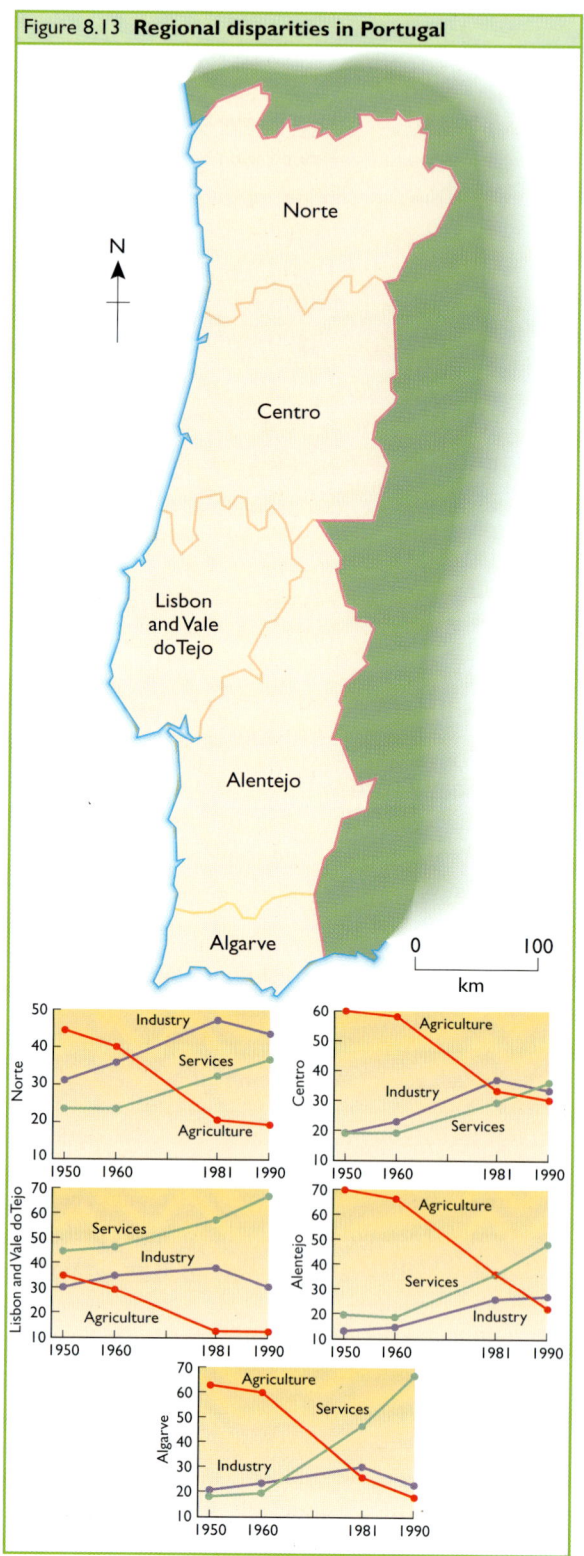

Figure 8.13 **Regional disparities in Portugal**

been transformed by foreign investment from a depressed shipbuilding and canning area into the hub of Portuguese industry. Foreign investors include **Ford Electronica**, **GM Delco Remi** and **Valnet Tractors** of Finland. The prime investment is the Esc 450 billion **Auto Europa** car plant, which is a joint venture between **Ford** and **VW**. The plant will produce 180,000 multi-purpose vehicles a year, accounting for 30% of Portugal's exports. Just-in-time production has encouraged local suppliers. By the end of 1992 over 30 suppliers had been given Q1 status by Ford VW. Q1 requires a reject rate of under 2% over a 12 month period.

Agricultural change

There is a clear **regional division** in Portuguese agriculture. Small family farms are found mainly in the north and much larger units in the cereals-dominated south. The problem is that family plots are too small to be competitive and the larger farmers are in need of modernisation. EU aid is being used to finance training programmes aimed at keeping young farmers on the land. Areas that have benefited include the region around **Lisbon** and along the **Tagus** river, where the proximity to a large market and easier access help to make farming more competitive. Other areas are facing competition from other EU producers and a reduction in prices and subsidies as a result of the reform of **CAP**. It is inevitable that the associated unemployment will be felt in the more backward **Norte** and **Centro** regions.

Regional effects

In most cases **Structural Funds** look likely to increase **regional disparities**. Little of the expenditure has gone to the poorest area of Portugal, the **Alentejo**, which has increased the core-periphery differences. This is partly because Alentejo's economy is based on agriculture, for which the **CAP** and **Agricultural Guidance Fund** have done very little. Indeed, increased competition has led to the decline of the agricultural sector in the interior. Most infrastructural development, and certainly the lion's share of inward investment, is restricted to the coastal strip between **Lisbon** and **Oporto**. Figure 8.13 shows the economic and sectoral difference of Portugal's regions. It can be argued that while the Structural Funds have led to a convergence of the Portuguese economy towards the European standard, very little has be done to combat the regional inequalities within the country itself.

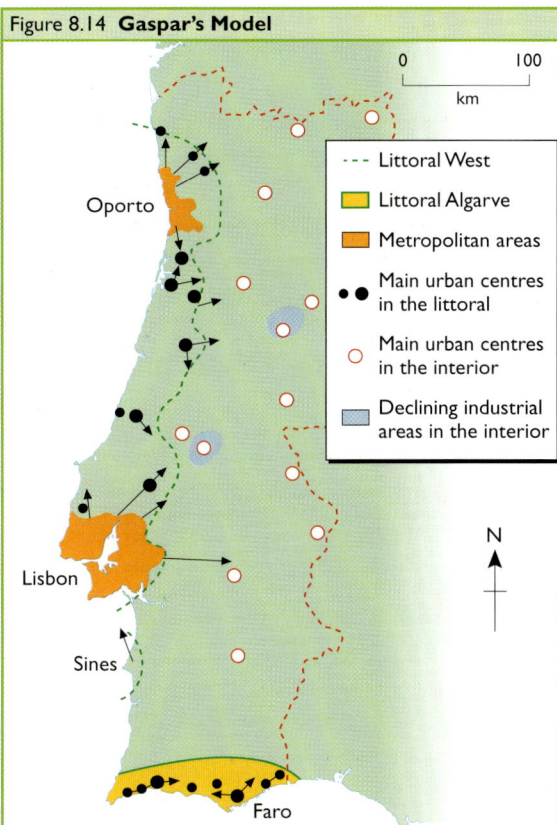

Figure 8.14 **Gaspar's Model**

Development and socio-economic relations in Portugal

The Portuguese geographer Jorge Gaspar has identified three main geographic dimensions to describe Portuguese space in terms of development and socio-economic relations: littoral-interior, north-south, and urban-rural (Gaspar, 1990).

Littoral (coastal) - interior division related mainly to the level of development, the littoral being more densely populated, more urbanised, more industrialised, with a more youthful population, attracting all major investment. The interior is, broadly speaking, the opposite.

The north-south divide has arisen through a long historical process, reflecting differences in social relations. The north is more religious, politically more conservative, and more rural.

The urban-rural dimension has introduced significant inversions of these general patterns, particularly in the distinction between littoral and interior. In littoral areas there is a diffusion of industrialisation and urbanisation representing an amelioration of the urban-rural distinction. In the interior the opposite process seems to be at work: urban centres and proto-centres are involved in a process of growth and restructuring, not only through the introduction of industry but also through development of service activities, related in turn to rural depopulation through migration.

SPAIN AND EU MEMBERSHIP

Spain's government made it clear from the start of the Maastricht talks that its top priority was to join the **'club of nations'** meeting the **convergence criteria** for economic and monetary union in time for the final phase of EMU. However, the current obsession with convergence, notably on inflation and currency stability, have left Spain without policy tools to spur domestic growth, especially the much needed investment in physical and human capital by domestically owned private businesses.

If Spain has stumbled trying to meet the requirements of convergence it has clearly benefited from joining the EU. At the beginning of June 1994 the EU approved **Structural Funds** worth ECU 26.3 billion for Spain between 1995 and 1999. This represented half of the Structural Funds approved for all the EU members and double the funding between 1989 and 1993.

Figure 8.15 Structural Funds and regional development

Infrastructure

• 1045 km of motorway are being co-financed by the ERDF within the framework of the Spanish motorway network programme.

• The Madrid-Seville high speed train (HST) is a major trans-regional project which was granted special ERDF assistance totalling ECU 784m. It runs through Ciudad Real and Cordoba and cuts journey time to three hours.

• Malaga airport has been modernised and equipped with a new international terminal.

• 18 medium-sized ports in Galicia have been modernised in addition to the ports of Gijón and Luarca in Asturias, Bilbao in the Basque country, Sanlucar in Andalucia, Tarragona in Cataluña and seven ports in the Canaries.

Water and power supply

• Water conveyance facilities at San Sebastian (Basque country) and regulation reservoirs and hydro-geological research in Castilla-La Mancha.

• Numerous projects to address power supplies including a thermo-electric coal residue burning power station in Asturias.

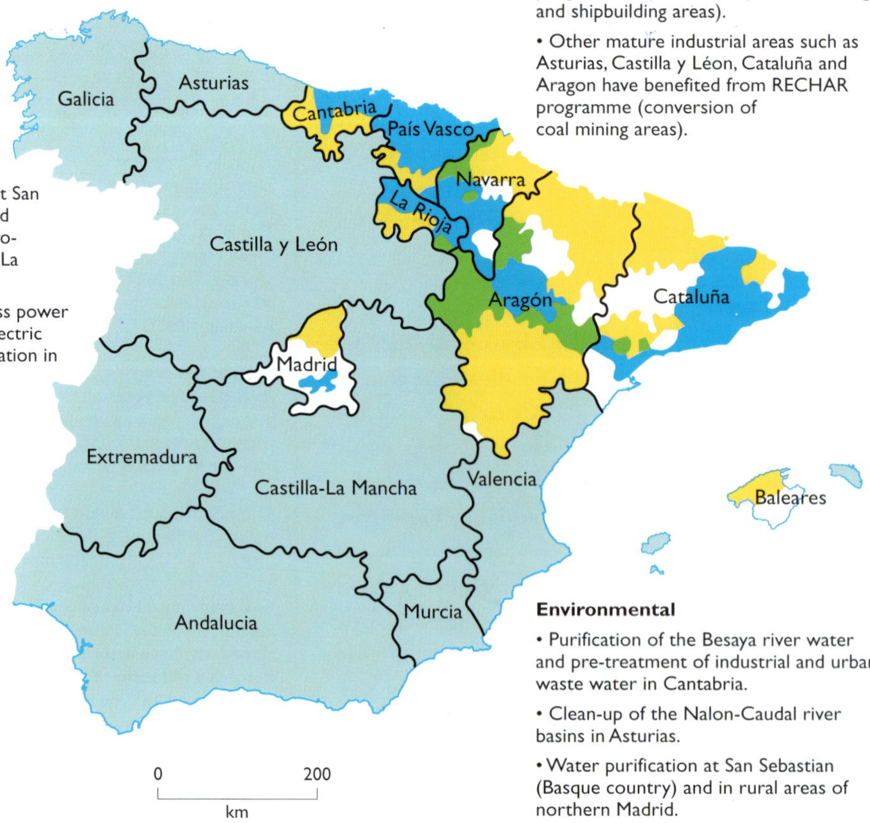

Industrial

• Industrial estates of varying sizes have been created in all less-developed regions and in declining industrial regions such as Madrid, the Basque country and Cataluña.

• Technology parks such as those at Malaga (Andalucia), Bilbao (Basque country), Vallés (Cataluña).

• In the Basque country declining industrial sectors have been supported under the community RESIDER and RENAVAL programmes (conversion of steel making and shipbuilding areas).

• Other mature industrial areas such as Asturias, Castilla y Léon, Cataluña and Aragon have benefited from RECHAR programme (conversion of coal mining areas).

Environmental

• Purification of the Besaya river water and pre-treatment of industrial and urban waste water in Cantabria.

• Clean-up of the Nalon-Caudal river basins in Asturias.

• Water purification at San Sebastian (Basque country) and in rural areas of northern Madrid.

Legend:
- Objective 1
- Objective 2
- Objective 5b
- Objective 2+5b

Regional problems

Spain joined the EU in 1986 barely ten years after its return to democracy. It ranks second among the member states in terms of area (505,000 km^2) and fifth in terms of population (40 million). There are **large regional disparities** within a country where less developed regions account for 76% of geographical area and contain 58% of the population. The worst areas include a series of declining mature industrial areas clustered between the centre and the north east and around Madrid and rural areas dependent on particularly vulnerable agricultural activities. Spain, therefore, combines the three types of area eligible for regional assistance from the Structural Funds (**Figure 8.15**). This entire group of regions suffers from serious structural handicaps and the rise in unemployment has affected a large number of young people and women entering an unbalanced labour market.

The Community Support Framework (CSF)

Figure 8.15 shows that the whole of Spain is covered by CSF finance. These are co-financed mainly by the member states and the Commission and are based on development plans submitted by the Spanish government to the EU:

1 **Objective 1** covers the nine autonomous states: Andalusia, Asturias, the Canaries, Castile-la Mancha, Castile-Léon, Valencia, Estremadura, Galicia, Murcia and two cities Centa and Melillia.

2 **Objective 2** covers all or part of 11 provinces in seven autonomous communities: Saragossa (Aragon), Cantabria, Barcelona, Gerona and Tarragora (Catalonia), Madrid, Navarre, La Rioja, Alava, Guipuzcoa and Viscaya (the Basque country).

3 **Objective 5b** partially covers 12 provinces in eight autonomous communities: Huesca, Saragossa and Teniel (Aragon), the Balearic Islands, Cantabria, Madrid, Gerona, Tarragona and Lérida (Catalonia), Navarre, La Rioja, Alava (Basque).

Regions and achievements

Between 1989 and 1993 Spain experienced an additional annual growth of about 3% compared with what would have been achieved without EU support. In 1993, GDP was 1.5% higher than it would have been without CSFs.

The rate of investment increased from 22.5% in 1988 to 26.5% in 1993 and overall support could help create between 80,000 and 105,000 jobs. Most of this new employment is centred on Madrid.

The future

Spain shares with Portugal the experience that membership of the EU is a bitter pill to swallow. Entry in 1986 saw a huge influx in foreign investment which means that now half of Spain's manufacturing capacity is in foreign hands. The motor industry, which is Spain's chief exporting sector, is entirely foreign controlled, but there is evidence that Spain is no longer the obvious place for investors seeking low wages, especially in comparison with former communist countries. Labour costs have increased and any advantage that still remains is offset by the gap in productivity.

Figure 8.16 Inward investment threatened

There seems to be a cooling of the heated rush to invest in Spain. In 1994 the following announcements were made:

1 Santana Motor factory, a subsidiary of Suzuki, in the Andalusian town of Linares, cut 1500 of the plant's 2400 jobs.

2 Kubota, another Japanese company, announced it was closing its EBR KUBOTA tractor factory outside Madrid with the loss of 300 jobs.

3 VW's Seat car subsidiary and Nissan's Motor Iberica operation, both high-profile acquisitions of the 1980s, are now going through drastic restructuring.

Two other examples illustrate the reasons for these concerns about the competitiveness of a Spanish location.

Gillette
After 27 years making razor-blades in Spain, Gillette decided to close its factory just outside Seville, where the workforce had already shrunk to less than 240 workers. Three factors explain the closure:

1 sacrifice of Seville to concentrate production in the UK and Germany;

2 the open European market which has reduced import barriers thus freeing the Spanish markets;

3 obsolete technology in Seville and the larger capacity of other plants.

Suzuki
By suspending payments to creditors in February 1994 and proposing drastic job cuts, Santana's Japanese management provoked an explosive reaction in the Linares region which is heavily dependent on the factory and its 30-odd supplier companies. Although the rationalisation has been softened by the socialist regional government, a complete closure seems likely. Three factors explain the cut-backs:

1 labour costs account for 20% of sales compared with 7% in Japan;

2 the plant is old and lacks investment in modern machinery;

3 the plant has made losses of between Pta 21bn and Pta 30bn during 1992-94.

Section D Exercises and recommended reading

EXERCISES

1. Using the data below, describe the changing pattern of employment in the Republic of Ireland and Northern Ireland. Write a reasoned account of the ways in which industrial and regional development in Ireland and Northern Ireland have differed and converged. Which do you think has greater prospects in the new Europe and why? [12]

Figure 8.17 Employment structures in Ireland and Northern Ireland, 1971-1990

	Ireland		Northern Ireland	
	1971	1990	1971	1990
Agriculture	26%	15%	11%	7%
Industry	30%	28%	42%	25%
Services	44%	57%	47%	68%

2. Using the information in this chapter and the table below, explain at least *three* ways in which countries have come to terms with peripherality. [13]

Figure 8.18 Coping with peripherality—(a) ignoring it (b) denying it

(a) 'Ireland offers a unique opportunity to the investor. It is a full member of the EU and provides high productivity and low inflation. It also offers:
—an ideal location for markets in the greater European area plus North America and the Middle East;
—a well educated English speaking workforce, with a plentiful supply of young graduate/technical personnel;
—duty free access to the EU market of 370 million consumers.'

Source: IDA, Ireland

(b) 'I have heard Northern Ireland referred to as a "peripheral location" and can only conclude that such sentiments are expressed by those who do not have any real experience in international business. We find that we can serve customers in North America and the Far East just as effectively as many suppliers closer to those markets. Geographic positioning does not make Northern Ireland peripheral; peripherality is a state of mind.'

Source: David Boulter, Chief Executive, Aircraft Furnishing

3. Write a report of not more than 500 words on why a high-tech firm should locate in any one country mentioned in this chapter rather than the other EU members. [13]

RECOMMENDED READING

There are many good books available on the countries covered in this chapter. These include Whyte (1990) *Interpreting Northern Ireland*, Carter & Parker (eds.) (1987) *Ireland: contemporary perspectives on a land and its people*, Brunt (1988) *Ireland* and McLaren *Dublin*. Recent articles include: 'Northern Ireland: Banners, bombs and babies' (1994) in *Geography Review*, vol. 8, no. 1, pp. 31-5, 'Inward Investment in Northern Ireland' in *Geographical*, September 1994, pp. 43-5, 'Northern Ireland catches up' in *Geographical*, November 1994, pp. 56-7, 'Portugal' (1995) in *Geofile* 269, 'What hope for Northern Ireland' (1995) in *Geofile* 270, 'Killarney National Park' in *Geographical*, May 1995, pp. 61-3, 'The Emerald Corridor' (1996) in *Geofile* 291.

Glossary

Agglomeration (external economies of scale) The savings made by an individual firm when it locates close to other firms thereby saving in terms of labour and infrastructural development.

Assisted Areas Areas with a poor location and/or economic structure which are eligible for government aid. They are a mixture of peripheral and economically marginal areas.

Branch plant economy A region that depends upon branch plants (of MNCs) for much of its employment. It is relatively unstable as (i) it lacks control over the decision making and (ii) many of the jobs are menial assembly positions.

Constrained location theory In large urban areas the high price of land, limited land availability, congestion and so on, cause many firms to shift to suburban and rural areas. Options in the urban areas are 'constrained'.

Counterurbanisation A recent trend involving the movement of people away from conurbations and large urban areas to small towns and rural areas.

Concentrated decentralisation The dispersal of a relatively large number of industries and people from large urban areas to a relatively limited number of small towns.

Decentralisation The movement of employment and population from a small number of large centres, frequently an over-populated core or capital city, to create a more dispersed pattern of employment. This is often done by governments as a form of regional readjustment.

Deindustrialisation The decline of manufacturing industry caused by, for example, increased mechanisation, foreign competition or exhaustion of resources. It has been most drastic in regions of heavy traditional industries.

Diffuse industrialisation The growth of firms and industries in areas formerly less developed, e.g. Highlands and Islands, East Anglia.

Export orientated industries Industries which earn foreign capital. Increasingly, they target global markets and have interests in many countries. (See also Import substitution industries.)

Externalities Side effects felt as a result of, for example, industrial development. Positive externalities include the generation of wealth and employment whereas negative externalities include rising land prices, congestion, pollution and so on.

Footloose industries Industries which are not restricted in their choice of location by requiring proximity to raw materials, labour supplies and so on. Modern high-tech industries are more likely to be footloose than traditional heavy industries which were constrained by access to raw materials and energy supplies.

Gross Domestic Product (GDP) is the total value of the goods and services produced within a country.

Gross National Product (GNP) is the sum total of goods and services produced within a country and earned overseas.

Greenfield site An industrial site found on the edge of a town on land which was not previously used for industry.

Growth pole A town or region in which there is a concentrated investment of resources in order to stimulate self-sustaining growth and to encourage multiplier effects.

Import substitution industries Industries designed to protect and stimulate a nation's own industries by imposing barriers on imports, e.g. Ireland in the 1930s and 1940s.

Industrial inertia The process whereby an industry remains in an area long after the initial attracting factors have disappeared, e.g. steel in Sheffield.

Inward investment Capital investment by a foreign company into a country, e.g. Nissan in the North East of England.

Just-in-case (Fordist) A method of production entailing large-scale, mass production techniques. Large stockpiles are kept of all component parts.

Just-in-time (Toyotist) A flexible pattern of organisation associated with Japanese firms. The central factory receives components from a wide variety of suppliers located within a few hours distance of the main factory. Speed of response is vital to meet demand.

Multinational companies (MNCs) An organisation having operations in a large number of countries. Generally, research and development is concentrated in growth areas of developed countries whereas assembly and production is located in developing countries or depressed regions. MNCs are economically and politically powerful.

Newly Industrialising Countries (NICs) Countries characterised by rapid economic growth rates, e.g. Malaysia, which have undergone rapid and successful industrialisation in recent decades.

Post industrial society Societies in which deindustrialisation has occurred and the majority of employment is in service industries.

Product life cycle A theory that manufactured products pass through a series of stages: development, maturity and standardisation. Each stage varies in terms of locational requirements.

Reindustrialisation The growth of high-technology and new industries following deindustrialisation.

Screwdriver industries Unskilled, low paid employment involving the routine assembly of component parts.

Small and medium sized enterprises (SMEs) Firms employing a relatively small number of workers. A region with a large number of SMEs would have more diversity than a region with the same numbers employed by MNCs.

Spatial division of labour The concentration of high paid, skilled work in the core of developed countries, compared with low paid, unskilled workers in the periphery or developing world.

Tertiarisation The rapid growth of the service sector in developed economies.

INDEX

A
Ageing population 4-6, 88, 143
Agglomeration 92
Agribusiness 24
Agriculture **19-35**, 44, 106, **107**, 108, 140-3, **148**
Aid **48-9**, 122-3, **148-9**
Alsace-Lorraine **69**
Amsterdam 44, 85
Appadural 101
Asfordby 61, 62
Attiki 90

B
Baden Wurtenburg 46, 47
Beef farming **24**
Behavioural approach 27
Belfast 72, 137
Belgium 2, 6, 74, 90
Benelux 7, 103
Berlin 8
Bilboa 44
Birchwood Science Park **42**
Birmingham 120-3
'Blue Europe' **37-8**
Bremen 44
Brewing 46
British Coal 61
British Steel **66-7**
Bruges 96
Brussels 44, 90, 93, 102

C
Cambridge 96
Canaries 90
Canary Wharf 92
CAP: see Common Agricultural Policy
Car industry 55-6, **74-9**, **123**, 125, 148, 151
Cassa per il Mezzogiorno **125**
Cereal farming 24
Central Belt (Scotland) **140**
Central place theory 112
Center Parcs 96
Channel Tunnel 7, 102, 140
Chemical industry **45**, 108, 110
Clark's sector model **10**
Clydeside 61
Clywd 143, 144
Coal 49, **60-3**, 128, 129, 130, 143, 147

Common Agricultural Policy 17, **28-31**, 32, 138
Common Fisheries Policy **37**
Communications 42, 84
Community Support Framework 151
Competitiveness 15, 66, 72, 120, 128, 146, 148
Consumer services **88**
Core areas **42**, 55, 148
Core-periphery **11**, 84, 138
Counter-urbanisation 94, 100, 110
Crofting 24
Cumulative causation **11**
Czech Republic 3, 76

D
Dairying **24**, 106
D'Avignon plan 64
Decentralisation 45, 90, 107, 112-113, 127
Deindustrialisation **44**, 63-4, 67, **88**, 92, 125, 128, 140
Denmark 2, 22, **24**, 46, 69, 90
Developing world **13**, 68, 107
Development areas **48**
Dicken 18
Diffusion **27**
Division of labour 44, 45, 46, 47, **52-4**, 55, 74, 76, 83, 147 (see also: Labour costs)
Dover and Deal 48
Dublin 103, 135
Dudley **94-5**

E
East Anglia 21-2, 24, 32, 42, 87, 92, 96, 116
East Germany 90
East Midlands 62, 90, 92, 116
Eastern Belt (Scotland) **140**
Eastern Europe 3, 10, 64, 108
Economies of scale 65
EEA: see European Economic Area
EFTA: see European Free Trade Association
Electronics 44, 45, **82**, 121, 140, 141, 143
Emerald Corridor **135**
Energy 42, 49, 60-3, 146, 147
Engineering 73, 108, 140

England 21, 32
Environmental issues 4, 5, **32-5**, 36, **57**, 62, 129, 141
Environmental technology **57**, 63, 129
Environmentally Sensitive Areas **34-5**
Epinal 69
ESAs: see Environmentally Sensitive Areas
EU: see European Union
EuroDisney 96
Europe **2**, 91
European Economic Area 3
European Free Trade Association 3
European Union 2, 3, 12, 13, 14, 15, 93, 135, 136, 137
Eutrophication 33
Externalities 63

F
Farm size **22**
Farming: see Agriculture
Financial aid 7, 8, 134, **136**, 142, **148**
Fishing **36-9**, 140, 141, 142
Flexible production 56
Flexible specialisation 126
Fordism **56**, **74**, 107
Four Cities Partnership **104-5**
Fragmentation 22
France 6-7, 22-3, 28, 64-5, 68, 69, 75, 83, 92
Frankfurt 44, 92, 101, 102
Friedmann 101

G
Gaspar's model **148**
GATT: see General Agreement on Tariffs and Trade
General Agreement on Tariffs and Trade 13, 31
Germany 2-3, 63-5, 74-5, **128-31**
Glasgow 45, 94, 118, 140
Glaxo 83
Global shift 14, 50, 64, **83**, 100, 103
Globalisation 12, 45, **55**, 75, 146
Global switching **83**
Glocalisation 55, **74-6**

Golden Triangle 8
Government aid 72, 147
Grampians 140, **141**
Greater Thessalonika **146-7**
Greece 7-8, 96-7, **146-7**
Green Heart 106, 110-3
Gross Domestic Product 14-5, **143**
Growth axes 112
Growth pole 66, 67, **79**, 112, 118, 125
Guaranteed prices 28

H
Hagerstrand 27
Hamburg 90, 92
Hardin 36
Harland and Wolff **72-3**
Helsinki 96
High-tech industry 44, 46-7, **82-7**, 129, 145
Highlands (of Scotland) 21-2, **24**, 42, **49**, **141**
Horticulture **106**
Hot banana **9**, 103
Hualon 72
Humphrys' model **53**
Hungary 3

I
Ile de France 6, 8, 85
Industrial districts 46, 126, 127
Innovation 46, **85**
Investment 16-7, 70-1, 117-8, 120-121, **123**, **134**, 140-1, 143-144, 148-9
Ireland 22, **24**, 134-5, **137-9**
Irish Box 38
Iron and steel 42, 45, 49-50, 61, **64-7**, 128, 130
Italy 4-6, 22-3, 64-9, **124-7**

J
Japan **12**, 36, 45, 55
Jaguar 123
Just-in-time **56**, **74**, 77, **78-9**, 148

K
Keeble 44, 98

L
Labour costs 6-7, 14-5, 72-3, 120-1 (see also: Division of labour)

La Défense 92
Lancashire 61, 68, 69
Land tenure **22**, 25
Land reform 125
Lazio 90
Less Favoured Areas 24, 142
Les Vosges **69**
LFAs: see Less Favoured Areas
Lisbon 85
Location 11, **42**
London 85, 87, 90, 92, 96, 100-3, 116, 134
Losch 50
Luxembourg 2, 6

M

Madrid 90, 151
Market gardening **24**, 106
Manchester 85, **103**, 118
Manufacturing industry 44, 46, **50-1**, **108**, **109**, **116**, 146-7
Massey's **53**
Massif Central 21, 22, 24
Mediterranean 21, **25**, 90, 96, 125
Merry Hill **94-5**
Merseyside 44, 74
Mezzogiorno 8, 22, 24, 25, 42, 49, 66, **124-5**, **127**
Midlands 61, 87, 96, 134
Milan 92, 124
Mini-mills **67**
MNCs: see Multinational Corporations
Mondeo 74
Multinational Corporations 12-3, 16-7, 44, 50, 68, 72, 82, 100-1, 117-8, 137-8
Multiplant firms **53**
Multiplier effect **11**, 123
Munich 85, 92
Myrdal 10-11

N

NAFTA: see North American Free Trade Agreement
Netherlands 22-24, 90, 92, **106-13**, 126
Newcastle 94
Newly Industrialising Countries 10, 12, 13, 147
NICs: see Newly Industrialising Countries
Nitrate pollution **33**

Nord—Pas de Calais 8, 60, 69
North American Free Trade Agreement 12
North East (UK) 8, 42, 45, 60, 61, 67, 70-1
North Sea oil 140, **141**, 142
North-South divide (Italy) **124**
North-South divide (UK) 48, 90, **116-9**
North West 42, 118
Northern Ireland **72-3**, 90, 92, 96, 116, 118, **134-6**, 137

O

Oil and gas 42, 60, 61
Oil industry 140, 141, 142, 143
Opportunity cost **7**
Objective 1 8, 30, 39, **49**
Objective 2 8, 39, **49**, 122, 126, 128
Objective 5b 30, 39, **49**, 122
Overfishing **37-9**
Oxford 92, 96

P

Paris 8, 44, 90, 92-94, 96, 100-3
Paris Basin 22, 24, 32
Periphery 42, 46, 55, **135-54**
Peripherality 48, 124, 138, 140
Polders 112
Population 137, 138-9
Poland 3, 76
Port Talbot 65, 67
Portugal **146**, **148-9**
Prato **126-7**
Problem regions 8-9
Producer services 88, 90, **92-3**, 100, 116
Product life cycle **52-4**, **83-4**
Productivity 15

Q

Quotas 13, 29, 64, 106

R

R&D: see Research and development
Randstad 42, 85, 101-3 **106-13**
Rationalisation **45**, 63-6, 70, 88, 118, 120, 125
Ravenscraig 45, 67, 76, 140
Recession 120
Regional development **124-7**

Regional inequalities **10-1**, **110-1**, **148-9**
Regional planning 48, 49, **112-3**, **122**, **124-9**
Reindustrialisation **44**, 108
Relocation 68
Renault 100
Research and development 82, **84**, 85, 117, 146
Restructuring **45**, **117**, 118, 146
Retailing **94-5**, 100
Rhone-Alpes 42, 69, 126
Rhone Valley 68
Rostow **10-11**
Rotterdam 102, 108, **110-1**, 113
Ruhr 42, 44, 49, 60, **63**, 68, **128-31**

S

Sambre-Meuse 60
Science parks 46, **83-4**
Schipol airport 111, 113
Scotland 116, 118, 134, 138, **140-2**,
Services **88-98**, 100-101, 103, 106, **109**, 116, 120, 125, 134, 140, 141, 146
Set-aside **34**
Sheep rearing **24**
Sheffield 44, 67
Shipbuilding **70-3**, 140, 148
Siemens **71**
Silicon Glen 140
Sinclair **27**
Single European Act 2
Single Market 4-5, 12-13, 118
Slovakia 3, 76
Soil erosion 32
Small and medium-sized enterprises **46-7**, 49, 72, **109**, 126
SMEs: see Small and medium-sized enterprises
SOMIVAL 24
South East (UK) **42**, **48**, 87, 90, 96, 116, 118, 126
South Wales 60, 61, 143-5
South West (UK) 90, 92, 96, 116-8
Soviet Union 12
Spain 44, 50, 56, 64-5, 150-1

Spatial division of labour: see Labour costs
Spatial margins 50-1
Specialisation 46-7, 102-3, 126-7, 146-7
Structural Funds **148-9**, **150-1**
Suburbanisation 94, 100, 110
Surpluses 29
Swan Hunter **70-1**

T

Taranto 66
Technopoles 46
Textile industry 61, **68-9**, **72-3**, 108, 124-7, 134, 137, 140-2, 146-8
Thames Gateway 8, 42, 49
Third Italy **125-7**
Tourism 88, 90, **96-7**, 100, 134, **136**, 141-5
Trading blocs 12-13
Tragedy of the Commons **36**
Transplant firms 12
Transport 7, 49-50, 65, 108, 111-2, 120, 135, 141-4
Triad 55, 74, **82**
Tyneside 44

U

UEA 12
Utrecht 44
UK 16-7, **48-9**, 64, 69-70, 74-5, 92-3, 126, 138, 140,
Unemployment 4-6, **138**, **143**, 146-8
Urban fringe 27
Urban Development Corporations 122
Urban intersections 113
USA 137

V

Van Veen 64
Vernon 52
Von Thunen 23, **26**

W

Wales 33-4, 44, 49, 134, **143-5**
Warwickshire Science Park 50, **87**, 121
Weber **50**, 65
West Midlands **78-9**, 90, 116, **120-3**
World Cities 90, **100-5**